HIGH PERFORMANCE LOGIC AND CIRCUITS FOR HIGH-SPEED ELECTRONIC SYSTEMS

SELECTED TOPICS IN ELECTRONICS AND SYSTEMS

Editor-in-Chief: **M. S. Shur**

*The complete list of the published volumes in the series can be found at
https://www.worldscientific.com/series/stes

Selected Topics in Electronics and Systems – Vol. 62

HIGH PERFORMANCE LOGIC AND CIRCUITS FOR HIGH-SPEED ELECTRONIC SYSTEMS

Editors

F. Jain
University of Connecticut, USA

C. Broadbridge
Southern Connecticut State University, USA

M. Gherasimova
University of Bridgeport, USA

H. Tang
Yale University, USA

World Scientific

NEW JERSEY · LONDON · SINGAPORE · BEIJING · SHANGHAI · HONG KONG · TAIPEI · CHENNAI · TOKYO

Published by

World Scientific Publishing Co. Pte. Ltd.

5 Toh Tuck Link, Singapore 596224

USA office: 27 Warren Street, Suite 401-402, Hackensack, NJ 07601

UK office: 57 Shelton Street, Covent Garden, London WC2H 9HE

British Library Cataloguing-in-Publication Data
A catalogue record for this book is available from the British Library.

Selected Topics in Electronics and Systems — Vol. 62
HIGH PERFORMANCE LOGIC AND CIRCUITS FOR HIGH-SPEED
ELECTRONIC SYSTEMS

Copyright © 2019 by World Scientific Publishing Co. Pte. Ltd.

ISBN 978-981-120-843-0

For any available supplementary material, please visit
https://www.worldscienti ic.com/worldscibooks/10.1142/11502#t=suppl

Preface

Further to the publication of "*High Performance Materials And Devices For High-Speed Electronic Systems*" in Selected Topics in Electronics and Systems – Volume 61, last August, this volume comprises selected research papers presented at the 27th annual symposium of the Connecticut Microelectronics and Optoelectronics Consortium (CMOC), organized by a team of seven academic institutions and about eighteen companies, on April 4, 2018 at the University of New Haven (Orange Campus). In this volume, we have consistently attracted authors from across the United States who have presented their work and have also contributed papers spanning a broad range from the area of modeling of strain and misfit dislocation densities, microwave absorption characteristics of nanocomposites, to X-ray diffraction studies.

Specific topics in this volume include:

- Modeling of strain relaxation and defect dynamics in buffer layers for semiconductor devices fabricated on lattice-mismatched substrates, which enables technology for advanced computer chips, multi-junction solar cells, detectors, and microwave transistors.

- Physical Unclonable Functions (PUFs) are probabilistic circuit primitives that extract randomness from the physical characteristics of devices. One of the papers outlines PUF design based on resistor and capacitor variations for low pass filters (LoPUF).

- Spatial wavefunction switching (SWS) FETs, which can process 2-bits per FET using CMOS-SWS logic, thus enabling multivalued logic (MVL) and compact DRAMs.

- Perimeter gated single-photon avalanche diode (PGSPAD). The applied voltage at the gate terminal modulates the electric field, making it uniform throughout the junction. This gating technique is an efficient method to prevent premature edge breakdown, one of the major problems in operating avalanche photodiodes diodes implemented in CMOS process.

In summary, papers presented in this volume cover various aspects of high performance logic and circuits for high-speed electronic systems.

Editors:

F. Jain (University of Connecticut)
C. Broadbridge (Southern Connecticut State University)
M. Gherasimova (University of Bridgeport)
H. Tang (Yale University)

Contents

Modeling and Control of a Multiphase Modular High-Frequency Converter/Inverter for Vehicle Applications

Kiarash Ahi

Alumnus of the University of Connecticut,
Storrs, CT 06269, USA
kiarash.ahi@uconn.edu

This paper presents a novel control algorithm for a modular high-frequency converter. This control algorithm is designed to achieve an effective frequency higher than the switching frequency on the passive elements. As a result, the ripple on the output is suppressed, and smaller capacitors can be used. In this work, the modular high-frequency converter is modeled by equivalent boost converters. Based on the equivalent models a control algorithm is developed. The accuracy of the algorithm has been verified by simulation results using PLECS in the MATLAB/Simulink environment.

Keywords: Modular converters; modular power electronics; boost converters; equivalent circuit models; multi-output buck-boost converters.

1. Introduction

In general, by taking advantage of high frequency switching technics, the required size, weight and cost of the passive storage elements of the circuit could be reduced greatly [1], [2]. The occupied volume and the imposed weight by passive storage elements of power electronic circuits are of important flaws of using of power electronics in the vehicles and portable applications. On the other hand, since the costs for the active semiconductor devices are steadily falling, increasing the frequency in order to decrease the sizes of the required passive elements could also lead to decrease in the cost of the converters [3]–[5]. Novel devices with higher frequency and at the same time higher power have been recently developed by use of wide-band semiconductor materials [6], [7].

In the future energy systems, fossil fuel vehicles will be replaced by electric vehicles, and thus electric vehicles need to compete with conventional vehicles in terms of costs, performance and reliability [8]–[11]. Since physical stresses and unpredicted accidents in vehicles are more than stationary applications, and a fault may happen in a vehicle while it is far from human communities or when it is in a critical mission, robustness is critical in vehicle applications. A vehicle has to be reliable enough to get to a town or possibly a repair shop in cases of mechanical and electrical faults [12]–[19]. In general, modular systems can be more fault-tolerable compared with non-modular ones [20]–[27]. Reliability makes utilization of modular systems highly appealing in vehicle applications

[28]–[34]. In addition, since modular systems are built by similar submodules, the cost and the time of production are reduced [35]. Moreover, maintenance and repairing of the modular systems are faster and more efficient [36], [37]. Due to the fact that the market in vehicle manufacturing is highly competitive in terms of costs and efficiency, the mention features make modular systems very desirable for vehicle industries. Since, unlike the previous works, the presented control strategy in this work is feedback-less, the resulted system is relatively robust and economical [38]. Overall, a system that combines implementation of high frequency technics and modular approaches is optimal for vehicle applications.

A circuit topology of a novel modular high frequency converter (MHFC) for vehicle applications has been introduced in literature [1], [39]. This topology with three submodules is shown in Fig. 1. As shown in this figure, the MHFC can be considered to be constructed of three parts: the battery side, input capacitors, and output H-bridges. The first and the main step towards controlling this MHFC is developing a control strategy for acquiring the desired voltages on the input capacitors of submodules. The proposed control strategy is sensor-less and consequently more robust and economical.

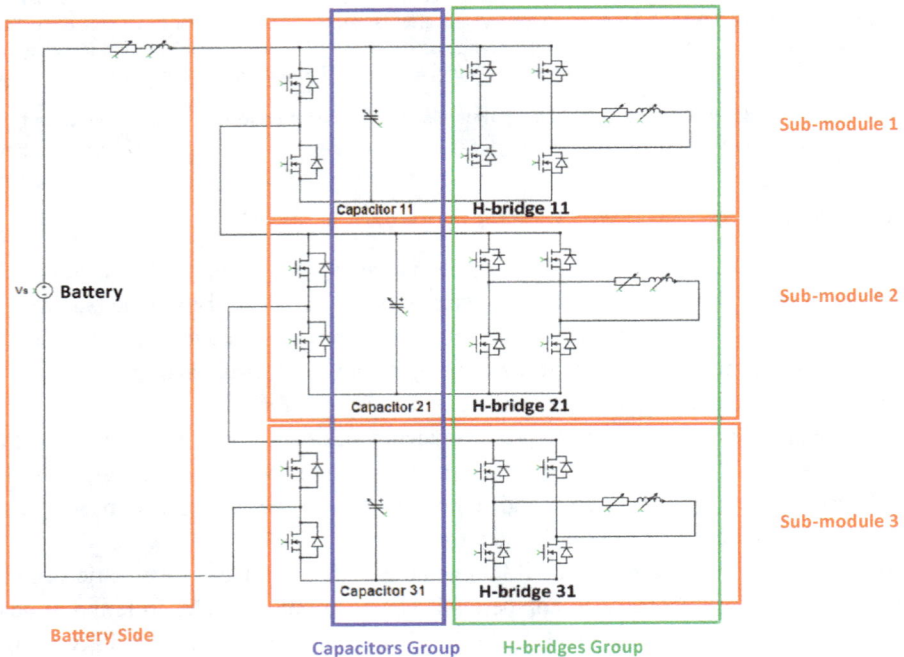

Fig. 1. Analyzed circuit topology of the MHFC with three sub-modules.

This paper is organized as follows: In Sec. 2, the MHFC is modeled by equivalent fundamental models and based on the equivalent models, control equations are derived. In Sec. 3, the control algorithm is developed to achieve the desired target voltages at the input capacitors of submodules. Target voltages could be either higher or lower than the battery voltage. In the Sec. 4 the accuracy of the algorithm has been verified by simulation results using PLECS in the MATLAB/Simulink environment. Section 5 is the conclusion.

2. Basic Fundamentals of Modelling MHFC by Boost Converters

The general topology of boost converter with N outputs is depicted in Fig. 2 [40].

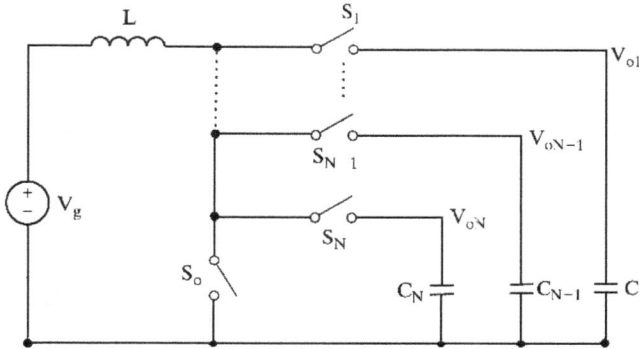

Fig. 2. Topology of a boost converter with N outputs.

Previously, some works have been done on modeling and controlling of multi-output boost converters [41], [42]. In previous works the converter works in discontinues conduction mode (DCM). By the beginning of each of the switching periods the current of the inductor rises from zero and before ending of that switching period it falls to zero again. In addition, only one output is fed during each of the switching periods. Each output is fed independently from the other outputs. Such converters are designed for low power applications such as in microprocessors. Although the proposed algorithms in the previous works are helpful in modeling the MHFC, it needs to be taken into account that the MHFC is designated to be utilized in power applications. Consequently, high current in the battery side is needed. In addition, the magnitude of the ripples in DCM are as high as the magnitude of the DC values of the currents and voltages [43]. It should be noted that ripples on the currents and voltages of storage components of this MHFC system are playing a major role in performance of this system: As much as the ripples are suppressed the sizes for storage elements of the circuit could be reduced. Based on these considerations, this MHFC is to be designed in continues conduction mode (CCM).

In general, a typical boost converter, during each switching period, is represented by two different circuits (modes of operation). The main idea is to obtain a particular DC current in the left-hand side of the circuit (battery side) and then obtain certain DC voltage on the capacitor by charging the capacitor via the current in certain portion of each of the switching period.

Assuming the source voltage is constant, in steady state, the variables on which the capacitor voltage of a boost converter depends are:

1. Capacitor's input current
2. Switching duty cycle

And the variables on which the inductor current depends are:

1. Switching duty cycle
2. Capacitor's voltage

Therefore, the capacitor voltage, the variable that is to be controlled, is depend on inductor current and both of these variables are depending on the switching duty cycle. Consequently, both the capacitor's voltage and inductor's current need to be controlled in a way that their required portion from duty cycle is satisfied. Also, for controlling the proposed modular converter, a procedure is to be developed in order to provide the required duty cycle for each of the capacitors while the required duty cycle for the inductor is being satisfied as well. Towards this aim, the proposed modular converter system could be modeled by two equivalent sub-models. First sub-model, sub-model-1, is an equivalent boost converter as seen from each of the terminals of submodules' capacitors. The second sub-model, sub-model-2, is another equivalent boost converter as seen from the terminals of the battery side in Figs. 1 and 2. These two sub-models comprise the equivalent mode of the MHFC together.

What should be taken into account is that in a boost converter the switching duty cycle which is being sensed by each inductor/capacitor (storage elements) together with the voltage which held across the terminals of that elements in each mode are determining the voltage/current of that element. In a boost converter, if the inductor senses mode-1 for k portion of the switching period, T, then the capacitor will sense mode-2 for (1-k) portion of T. The interesting fact is that in an equivalent modeled system of this MHFC the equivalent modeled capacitor and the equivalent modeled inductor sense different switching frequencies. The switching frequency which is applied to the capacitors is higher than the switching frequency which applies to the inductor. The inductor in sub-model-2 represents the actual inductor of the MHFC and switching frequency which is being applied to this inductor is the actual switching frequency which is applied to the actual inductor of the MHFC. In addition, since the battery and the inductor are paralleled with only one switch in the modeled boost converter, sub-model-2, while in the real system this is not the case, the switch in the modeled system would be a virtually modeled switch. In the same manner the capacitor in sub-model-1, represents each of the three input capacitors of MHFC, in addition the switch in this equivalent model represents input switch of the respective sub-module and thus the switching frequency and capacitor are both the actual elements of the MHFC. In summary, in an equivalent modeled system of this MHFC, it is possible to have two equivalent modeled boost converters in which the parameters are as follows:

1-Parameters of sub-model-1:

$$\begin{cases} \text{Switching duty cycle} = k^c = k \\ \text{Switching period} = T^c = T \end{cases}$$

2-Parameters of sub-model-2:

$$\begin{cases} \text{Switching duty cycle} = k^i \\ \text{Switching period} = T^i \end{cases}$$

It should be taken into account that in sub-model-1 the switch represents the input switch of the corresponding submodule while in the sub-model-2 the switch represents a virtual switch. In addition, the capacitor in sub-model-1 represents an actual capacitor of any of the three submodules while in the sub-model-2 the capacitor is a virtual capacitor in the model.

2.1. *Developing the equivalent modeled boost converter of the MHFC as seen from the terminals of the battery side*

For developing the equivalent model of the proposed modular system as seen from the terminals of the battery side, sub-model-2, it should be noted that the current of inductor in the first portion of the switching period needs be rising while in another portion it should be falling. In other words, the voltage across the inductor should be positive in the first portion of the switching period and negative in the second portion of the switching period. It is not necessary to short circuit the inductor with battery in the first portion of switching period. For obtaining this requirement in the proposed modular system, in first portion of the switching period, mode-1 of operation, less number of capacitors (submodules) needs to be in series compared to the second portion of the switching period, mode-2 of operation. This idea could be implemented to the MHFC with arbitrary submodules. In continue implementation of the above idea to a system with 3 submodules is investigated.

Assuming the battery is 180V and the required voltage on the capacitors 100V as considered in the prior works [1]. The capacitor voltage is lower than the source voltage in this case. For the given values, for constructing mode-1 of operation, only one capacitor is to be in series with the battery side while for constructing mode-2 of operation two capacitors is to be in series to the battery side. By taking each capacitor voltage as V_c, the equivalent model of the proposed MHFC as seen from the terminals of the battery side, sub-model-2, could be obtained. The scheme of this sub-model is depicted in Fig. 3(a).

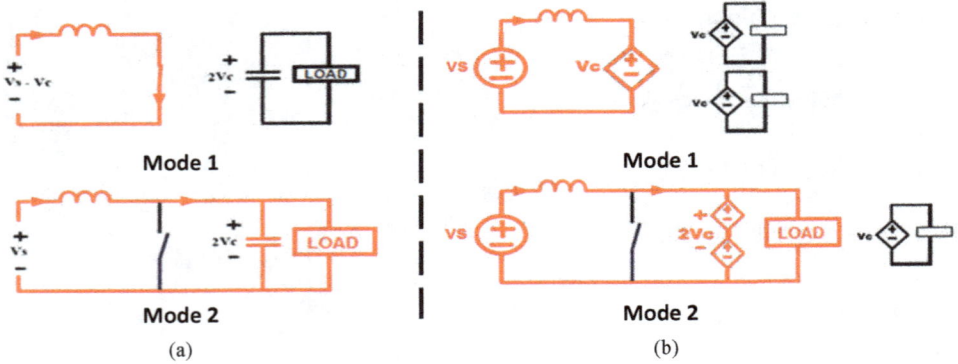

Fig. 3. (a) An equivalent circuit of the proposed modular converter system as seen from the terminals of the battery side; (b) submodules' input capacitors are modeled as dependent voltage sources.

For obtaining rising current in mode-1 of operation:

$$0 < Vs - V_c \Rightarrow V_s < V_c \tag{1}$$

For obtaining falling current in mode-2 of operation:

$$V_s < 2V_c \Rightarrow \frac{1}{2}V_s < V_c \tag{2}$$

Combining the last two equations yields:

$$\frac{1}{2}V_s < V_c < Vs \tag{3}$$

For the equivalent modeled boost converter of Fig. 3, the following calculations could be done.

In mode-1:

$$k^i T^i = \frac{\Delta IL}{V_s - V_c} \tag{4}$$

In mode-2:

$$(1 - k^i)T^i = \frac{\Delta IL}{2V_c - V_s} \tag{5}$$

Solving the above equation for k^i yields:

$$k^i = \frac{2V_c - V_s}{V_c} \tag{6}$$

Equation (6) shows that the equivalent duty cycle which is seen by the inductor, k^i, controls the voltages of the capacitors. Accordingly, in the equivalent circuit as seen by the terminals of battery side, submodules group can be considered as dependent voltage source, V_c, as shown in Fig. 3(b).

2.2. *Developing the equivalent modeled boost converter of the MHFC as seen from the terminals of each of the submodules' capacitors*

In this equivalent model each sub-module (and consequently its input capacitor) senses mode-1 of operation when its input switch (upper switch of the input half bridges of the sub-module) is off and it senses mode-2 of operation when its input switch is on. The input switches of the other two submodules may be on or off independently. According to the mentioned facts, by taking the inductor (common bus) current as I_b, the equivalent model of the proposed modular system as seen from input terminals of each of the submodules is obtained as depicted in Fig. 4(a).

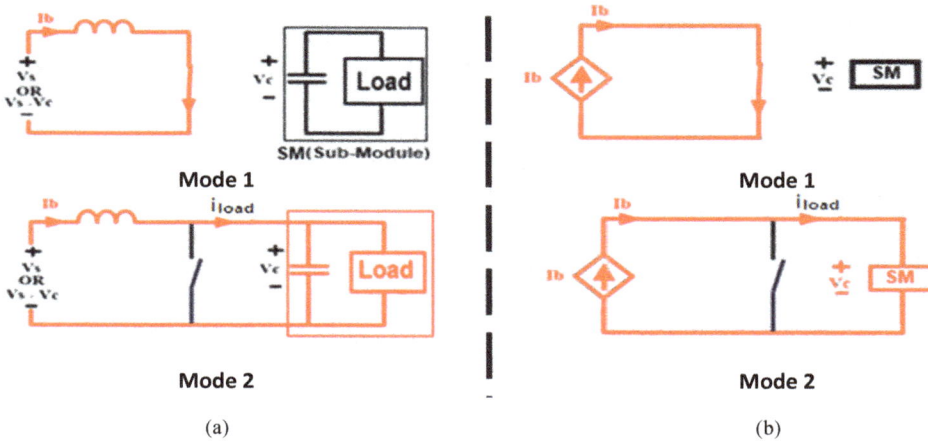

Fig. 4. (a) An equivalent circuit of the proposed modular converter system as seen from the terminals of each of the sub-modules' capacitors; (b) current of the battery side is modeled as dependent current source.

Input energy to each sub-module (sm) could be derived by:

$$E_{in,sm} = V_c I_b (1 - k^c) T \tag{7}$$

And the output energy:

$$E_{out,sm} = \frac{V_c^2}{Z_{load}} T \tag{8}$$

According to the conservation of energy:

$$E_{in,sm} = E_{out,sm} \tag{9}$$

And then:

$$I_b = \frac{V_c}{Z_{load}(1-k^c)} \tag{10}$$

Equation (10) shows that the equivalent duty cycle which is seen by the capacitor, kc, controls the current of the common bus (battery current). Accordingly, in the equivalent circuit, the battery side of the MHFC could be considered as a dependent current source, I_b, as shown in Fig. 4(b). In addition, equation (10) can be utilized for calculating the common bus current theoretically.

3. Developing a Control Strategy Aimed at Targeting the Commanded Voltage on Submodules' Input Capacitors

In the proposed MHFC system the switches are in parallel with capacitors, therefore capacitors sense the real switching frequency, while it is possible for the inductor to sense an equivalent effective switching frequency and consequently kc is the real switching duty

For controlling the system, two switching signals may overlaps in a way that the desired ki (which is determined by equation (6)) is shaped. The strategy needs to be able to produce ki for every desired value between 0 and 1. Consequently, a switching pattern in which the widths of switching signals are between the two above states may produce the required 0<ki<1.

In summary, the commanded voltage to the system, Vc*, determines the virtual switching duty cycle, ki, through equation (6), then a proper algorithm determines real switching duty cycle, kc, and corresponding switching patterns.

For shaping the required ki, the following pulse pattern could be utilized as switching signals:

$$S_j(t) = \sum_{n=0}^{\infty} \left(u\left(t-(n+\frac{j}{N})T^c\right) - u\left(t-\frac{1}{N}T^c(1+k^i+N(n+\frac{j}{N}))\right) \right) \tag{11}$$

Where $S_j(t)$ is switching pulse of the input switch (upper switch of the input half-bridge) of sub-module j, N is the number of submodules, $(T^c)^{-1}$ is switching frequency and k^i is the required duty cycle as seen from the terminals of the battery side and is determined through equation (6) by substituting the commanded voltage as V_c.

The capacitors are in parallel with the switches and consequently they see the real switching operation, while the inductor sees an effective switching pattern. Base on the above descriptions and the model which has been developed in Fig. 4, the following effective switching pattern is applied to the inductor:

$$S^i(t) = \left(\sum_{i=0}^{N} S_n(t) \right) - 1 \tag{12}$$

Solving the above equation yields:

$$S^i(t) = S_1\left(\frac{t}{3}\right)$$

(13)

In this method, the switching frequency as seen by the inductor is N times larger the physical switching frequency. This leads to the need for an inductor which is N times smaller.

For the case of a modular converter system of this type with three submodules, equation (11) yields:

$$S_j(t) = \sum_{n=0}^{\infty}\left(u\left(t-\left(n+\frac{j}{3}\right)T^c\right) - u\left(t - \frac{1}{3}T^c\left(1+k^i+N\left(n+\frac{j}{3}\right)\right)\right)\right)$$

(14)

And then:

$$S_1(t) = \sum_{n=0}^{\infty}\left(u(t-nT^c) - u\left(t - \frac{1}{3}T^c\left(1+k^i+3n\right)\right)\right)$$

(15)

$$S_2(t) = \sum_{n=0}^{\infty}\left(u\left(t-\left(n+\frac{1}{3}\right)T^c\right) - u\left(t - \frac{1}{3}T^c\left(1+k^i+3\left(n+\frac{1}{3}\right)\right)\right)\right)$$

(16)

$$S_3(t) = \sum_{n=0}^{\infty}\left(u\left(t-\left(n+\frac{2}{3}\right)T^c\right) - u\left(t - \frac{1}{3}T^c\left(1+k^i+3\left(n+\frac{2}{3}\right)\right)\right)\right)$$

(17)

In this case switching frequency as sense by inductor is three times higher than the real switching frequency.   □

4. The Control Strategy for Obtaining Capacitors' Voltages Higher than the Source Voltage

In Sec. 2 a strategy was developed towards obtaining capacitors' voltages lower than the DC voltage source, within the range which equation (3) indicates. For implementation of that control strategy, switching signals of equation (11) is applied to the submodules switches of the MHFC where k^i is determined by equation (6). In the same way, another control strategy could be devised for obtaining voltages on the capacitors to be higher than the DC source voltage, i.e.:

$$V_s < V_c$$

(18)

After assumptions and calculations similar to those of the previous section, the following equations would be derived for implementation of this control strategy:

$$k^i = \frac{V_c - V_s}{V_c}$$

(19)

$$S_i(t) = \sum_{n=0}^{\infty}\left(u(t-n\times T^c) - u\left(t-(n+1)\times T^c + \frac{k^i T^c}{3} - \frac{i\times T^c}{N}\right)\right) \qquad (20)$$

It should be noted that in this case all the submodules' input switches are on in mode-1 and only one sub-module switch is on in mode-2 of operation.

5. Simulation Results

Substituting k^i from equation (6) into equations (14–16) yields the wave forms (switching pulses) that are shown in Fig. 5.

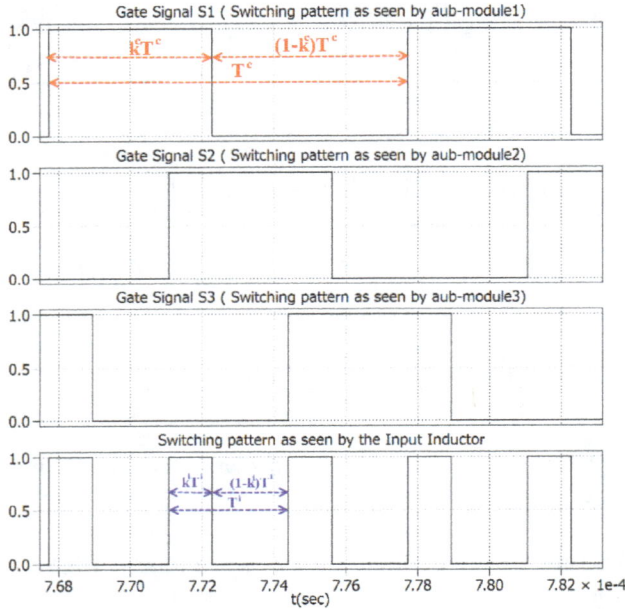

Fig. 5. Switching pulses for obtaining capacitors' voltages lower than voltage source (Equations (15–17)).

The effective switching pattern which is applied to the virtual switch of Fig. 4 is also depicted in Fig. 5. This pattern is what that had been presented theoretically by equation (13). The frequency of this effective switching pattern is 3 times higher than the physical switching frequency of the system's transistors; in this case the switching frequency is 100 KHz while an effective switching pattern with a frequency of as 300 KHz is applied to the inductor. Battery voltage is constant value of 180V. Figure 6 shows that by applying the proposed algorithm, the commanded (reference) voltage, V_r, has been acquired on input capacitors of submodules. In addition, for testing dynamic response of the system, in Fig. 6, the commanded voltage is changed from 130 to 140, the submodules' input capacitors targeted the commanded voltage with a settling time [1] which is less than 10 milliseconds and an overshoot which is less than 2.5%.

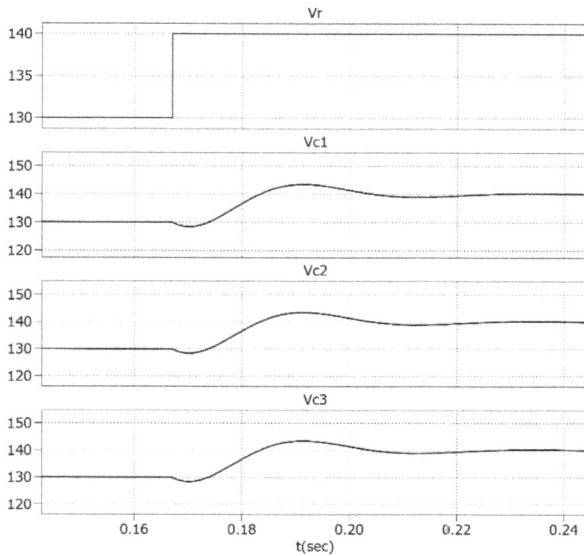

Fig. 6. Reference voltage and targeted voltages on sub-modules' capacitors (source voltage is Vs=180V).

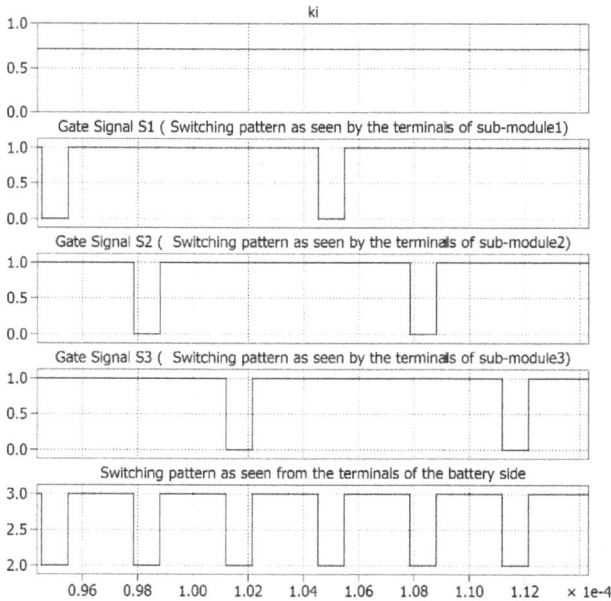

Fig. 7. Switching pulses for obtaining capacitors' voltages higher than voltage source (Equations (20)).

Figure 7 illustrates switching signals of equation (20) for obtaining desired voltages of 630V on the capacitors; and Fig. 8 confirms the accuracy of the strategy of Sec. 4 for achieving higher voltages than that of the source on the capacitors.

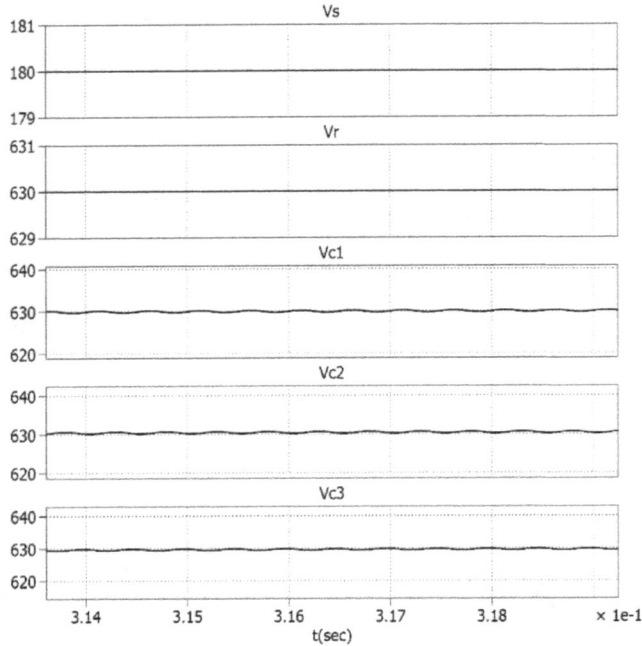

Fig. 8. Reference voltage and targeted voltages on sub-modules' capacitors (source voltage is Vs=180V).

6. Conclusion

In this paper a novel equivalent model for the MHFC is developed. This equivalent model is used for deriving a global control algorithm towards controlling this MHFC. It was showed that by taking advantage of this control strategy, not only the system targeted the reference voltage accurately but also an effective switching pattern with higher effective switching frequency is achieved on the input inductor of the MHFC. This leads to the need for a smaller inductor. The importance of this technique is due to the fact that the effective switching frequency is several times higher than the frequency limits of the switching transistors. Since, unlike the previous works, the presented control strategy is feedback-less, the resulted system is relatively robust and economical. The simulation results verify the accuracy of the control strategy.

Acknowledgments

The presented work in this paper is mostly based on the Master's thesis of the author that was done in Leibniz University of Hannover. The author would like to express his gratitude to his advisors Professor Axel Mertens and Dr. Arvid Merkert. The author would also like to thank Dr. Alireza Jahangiri, Dr. Ali Abedini, for their guidance on the topic.

References

[1] L. Lambertz, R. Marquardt, and A. Mayer, "Modular converter systems for vehicle applications," in *Emobility – Electrical Power Train, 2010*, 2010, pp. 1–6.

[2] V. Samavatian, S. Member, and A. Radan, "A High Efficiency Input / Output Magnetically Coupled Interleaved Buck – Boost Converter With Low Internal Oscillation for Fuel-Cell Applications: CCM Steady-State Analysis," *IEEE Trans. Ind. Electron.*, vol. 62, no. 9, pp. 5560–5568, Sep. 2015.

[3] M. Schulz, L. Lambertz, and R. Marquardt, "Dimensioning of Modular High Frequency converter for drives," in *2013 IEEE ECCE Asia Downunder*, 2013, pp. 675–680.

[4] S. Saridakis, E. Koutroulis, and F. Blaabjerg, "Optimization of SiC-Based H5 and Conergy-NPC Transformerless PV Inverters," *IEEE J. Emerg. Sel. Top. Power Electron.*, vol. 3, no. 2, pp. 555–567, Jun. 2015.

[5] B. B. Hu and M. C. Nuss, "Imaging with terahertz waves," *Opt. Lett.*, vol. 20, no. 16, p. 1716, Aug. 1995.

[6] K. Ahi, "Review of GaN-based devices for terahertz operation," *Opt. Eng.*, vol. 56, no. 9, 2017.

[7] K. Ahi and M. Anwar, "A survey on GaN-based devices for terahertz photonics," in *Proceedings of SPIE – The International Society for Optical Engineering*, 2016, vol. 9957.

[8] S. Shajari, M. R. Norouzi, A. Abedini, and K. Ahi, "Coordinate control of TCSC based GA controller with PSS for stability improving in single machine system," in *2011 10th International Conference on Environment and Electrical Engineering*, 2011, pp. 1–4.

[9] K. Ahi, "Developing next generation of electric grids for fulfilling deficiencies of conventional grids in supporting today's requirements," in *2011 International Conference on Power Engineering, Energy and Electrical Drives*, 2011, pp. 1–7.

[10] M. Pakkhesal, K. Ahi, S. Bathaee, and S. M. M. Mohebi, "Multi-criterion management of AMI residual subscribers in power shortages," in *2011 10th International Conference on Environment and Electrical Engineering*, 2011, pp. 1–4.

[11] H. Nasiri, A. Radan, A. Ghayebloo, and K. Ahi, "Dynamic modeling and simulation of transmotor based series-parallel HEV applied to Toyota Prius 2004," in *2011 10th International Conference on Environment and Electrical Engineering*, 2011, pp. 1–4.

[12] M. E. Y. G. A. Emadi, *Modern electric, hybrid electric, and fuel cell vehicles : fundamentals, theory, and design*. Boca Raton, Fla. [u.a.]: CRC Press, 2010.

[13] Sangshin Kwak, Taehyung Kim, and Gwangmin Park. "Phase-Redundant-Based Reliable Direct AC/AC Converter Drive for Series Hybrid Off-Highway Heavy Electric Vehicles," *IEEE Trans. Veh. Technol.*, vol. 59, no. 6, pp. 2674–2688, Jul. 2010.

[14] S. A. K. M. Niapour and M. Amirabadi, "A highly reliable single-stage converter for Electric Vehicle applications," in *2016 IEEE Applied Power Electronics Conference and Exposition (APEC)*, 2016, pp. 3704–3711.

[15] S. M. Mousavi Sangdehi, S. Hamidifar, and N. C. Kar, "A Novel Bidirectional DC/AC Stacked Matrix Converter Design for Electrified Vehicle Applications," *IEEE Trans. Veh. Technol.*, vol. 63, no. 7, pp. 3038–3050, Sep. 2014.

[16] J. L. F. Daya, P. Sanjeevikumar, F. Blaabjerg, P. W. Wheeler, and J. O. Ojo, "Implementation of Wavelet-Based Robust Differential Control for Electric Vehicle Application," *IEEE Trans. Power Electron.*, vol. 30, no. 12, pp. 6510–6513, Dec. 2015.

[17] F. M. Ibanez, J. M. Echeverria, D. Astigarraga, and L. Fontan, "Soft-switching forward DC–DC converter using a continuous current mode for electric vehicle applications," *IET Power Electron.*, vol. 8, no. 10, pp. 1978–1986, Oct. 2015.

[18] K. Kandasamy, M. Vilathgamuwa, and K. J. Tseng, "Inter-module state-of-charge balancing and fault-tolerant operation of cascaded H-bridge converter using multi-dimensional modulation for electric vehicle application," *IET Power Electron.*, vol. 8, no. 10, pp. 1912–1919, Oct. 2015.

[19] K. Ahi, "Lithography, Spectroscopy, and Super Resolution Terahertz Imaging for Quality Assurance and Authentication," *Dr. Diss.*, Apr. 2017.

[20] K. Ahi, S. Shahbazmohamadi, and N. Asadizanjani, "Quality control and authentication of packaged integrated circuits using enhanced-spatial-resolution terahertz time-domain spectroscopy and imaging," *Optics and Lasers in Engineering*, Jul. 2017.

[21] R. Picas, J. Zaragoza, J. Pou, and S. Ceballos, "Reliable Modular Multilevel Converter Fault Detection With Redundant Voltage Sensor," *IEEE Trans. Power Electron.*, vol. 32, no. 1, pp. 39–51, Jan. 2017.

[22] M. Abdelsalam, M. Marei, S. Tennakoon, and A. Griffiths, "Capacitor voltage balancing strategy based on sub-module capacitor voltage estimation for modular multilevel converters," *CSEE J. Power Energy Syst.*, vol. 2, no. 1, pp. 65–73, Mar. 2016.

[23] A. Jahangiri and A. Radan, "Indirect matrix converter with unity voltage transfer ratio for AC to AC power conversion," *Electr. Power Syst. Res.*, vol. 96, pp. 157–169, 2013.

[24] A. Jahangiri and A. Radan, "A Simplified and Fast DSP-CPLD-Based Implementation Method of Space Vector Modulation Applied in Indirect Matrix Converters," *EPE J.*, vol. 23, no. 3, pp. 22–29, Sep. 2013.

[25] K. Ahi, A. Rivera, and M. Anwar, "Encrypted Electron Beam Lithography Nano-Signatures for Authentication," *Int. J. High Speed Electron. Syst.*, vol. 26, no. 03, p. 1740017, Sep. 2017.

[26] K. Ahi, N. Asadizanjani, M. Tehranipoor, and M. Anwar, "Authentication of electronic components by time domain THz Techniques," in *Connecticut Symposium on Microelectronics & Optoelectronics*, 2015.

[27] K. Ahi, N. Asadizanjani, S. Shahbazmohamadi, M. Tehranipoor, and M. Anwar, "THZ Techniques: A Promising Platform for Authentication of Electronic Components," in *CHASE Conference on Trustworthy Systems and Supply Chain Assurance*, 2015.

[28] A. Mayer, M. Schulz, C. Rolff, and R. Marquardt, "Fault tolerant operation of the Modular High Frequency Converter," in *2015 9th International Conference on Power Electronics and ECCE Asia (ICPE-ECCE Asia)*, 2015, pp. 2917–2924.

[29] S. M. Goetz, A. V. Peterchev, and T. Weyh, "Modular Multilevel Converter With Series and Parallel Module Connectivity: Topology and Control," *IEEE Trans. Power Electron.*, vol. 30, no. 1, pp. 203–215, Jan. 2015.

[30] Y. Zhao, C. Li, M. Zhao, and S. Xu, "Model design on emergency power supply of electric vehicle," *Math. Probl. Eng.*, 2017.

[31] F. H. Khan, L. M. Tolbert, and W. E. Webb, "Hybrid Electric Vehicle Power Management Solutions Based on Isolated and Nonisolated Configurations of Multilevel Modular Capacitor-Clamped Converter," *IEEE Trans. Ind. Electron.*, vol. 56, no. 8, pp. 3079–3095, Aug. 2009.

[32] M. Quraan, P. Tricoli, S. D Arco, and L. Piegari, "Efficiency Assessment of Modular Multilevel Converters for Battery Electric Vehicles," *IEEE Trans. Power Electron.*, vol. 32, no. 3, pp. 2041–2051, Mar. 2017.

[33] A. Jahangiri, A. Radan, and M. Haghshenas, "Stationary frame current regulation of Indirect Matrix Converter with power factor correction under distorted supply voltage," in *2010 1st Power Electronic & Drive Systems & Technologies Conference (PEDSTC)*, 2010, pp. 106–110.

[34] K. Varesi, A. Radan, A. Ghayebloo, and M. R. Nikzad, "An Efficient Methodology Proposed For Deciding About the Number of Battery Modules In Hybrid Electric Vehicles," *Autom. – J. Control. Meas. Electron. Comput. Commun.*, vol. 57, no. 1, Jul. 2016.

[35] K. Ahi, "Control and Design of a Modular Converter System for Vehicle Applications," Leibniz Universität Hannover, 2012.

[36] J. D. Van Wyk and F. C. Lee, "Power electronics technology at the dawn of the new millenium-status and future," in *Power Electronics Specialists Conference, 1999. PESC 99. 30th Annual IEEE*, 1999, vol. 1, pp. 3–12.

[37] T. Ericsen and A. Tucker, "Power Electronics Building Blocks and potential power modulator applications," in *Power Modulator Symposium, 1998. Conference Record of the 1998 Twenty-Third International*, 1998, pp. 12–15.

[38] A. Mayer, C. Rolff, and R. Marquardt, "Control concept and stability considerations of the Modular High Frequency Converter," in *2014 16th European Conference on Power Electronics and Applications*, 2014, pp. 1–10.

[39] R. M. A. Mayer L. Lambertz, "Control concept of the modular high frequency – converter for vehicle applications," *PCIM Europe*. Nuremberg, Germany.

[40] M. H. Rashid, *Power electronics handbook: devices, circuits, and applications*, 3rd ed. Amsterdam [u.a.]: Elsevier/BH, Butterworth-Heinemann, 2011.

[41] D. Maksimovic and S. Cuk, "A unified analysis of PWM converters in discontinuous modes," *Power Electron. IEEE Trans.*, vol. 6, no. 3, pp. 476–490, 1991.

[42] M. Dongsheng, K. Wing-Hung, T. Chi-Ying, and P. K. T. Mok, "A 1.8 V single-inductor dual-output switching converter for power reduction techniques," in *VLSI Circuits, 2001. Digest of Technical Papers. 2001 Symposium on*, 2001, pp. 137–140.

[43] L. M. P. Fanjul, "Design considerations for dc-dc converters in fuel cell systems," Texas A&M University.

Macromodel of G^{4}FET Enabling Fast and Reliable SPICE Simulation for Innovative Circuit Applications

Md Sakib Hasan*, Samira Shamsir, Mst Shamim Ara Shawkat,
Frances Garcia, Syed K. Islam and Garrett S. Rose

*Department of Electrical Engineering and Computer Science,
The University of Tennessee, Knoxville, TN 37996, USA*
*mhasan4@utk.edu

A macromodel of silicon-on-insulator (SOI) four-gate transistor (G^{4}FET) is presented in this paper to aid circuit designers to explore innovative applications circuit with this multi-gate transistor. A number of works based on analytical solution, numerical simulation and experimental results of G^{4}FET have been previously reported. However, designing new interesting circuits with G^{4}FETs requires a SPICE model that will work sufficiently well throughout the desired operating regions. Although it is theoretically possible to solve coupled non-linear differential equations to explore different operating conditions, this will take an excessive amount of time making it unsuitable for useful circuit design. Therefore, a macromodel approach is adopted in this work to provide a reasonably fast and accurate circuit simulation. G^{4}FET combines the functionality of MOSFET and JFET devices which already have robust, fast and reliable SPICE models. A macromodel approach which is capable of combining these existing models including the interactions between multiple gates will be beneficial for any circuit designer. The feasibility of the macromodel is justified by simulating several analog and digital circuits and comparing against available experimental results.

Keywords: Silicon-on-insulator (SOI); multiple-gate transistor; G^{4}FET; semiconductor device models; SPICE; macromodel.

1. Introduction

There have been amazing technological advancements in semiconductor industries over the last several decades following Moore's law [1]. However, scaling down bulk silicon devices has become increasingly difficult and is now facing some fundamental physical limits. Some of the non-idealities such as subthreshold conduction, gate oxide leakage, threshold voltage roll-off due to DIBL (drain induced barrier lowering), reduced carrier mobility due to impurity scattering by increased doping concentration, slowing down of switching time scaling due to more dominant role played by interconnect capacitance, reverse-biased junction leakage etc. can no longer be ignored. To solve the problems associated with bulk silicon scaling and enable the semiconductor industry to extend Moore's law in the foreseeable future, researchers have been searching for new process technologies. One of the promising candidates is silicon-on-insulator (SOI) technology with many advantages such as ideal device isolation, reduced parasitic capacitance, excellent sub-threshold slope, elimination of latch up, increased switching speed, radiation hardness, reduced leakage current etc. [2]. A recent SOI device with multiple gates that has been used for several innovative circuit designs is G^{4}FET [3, 4]. The name G^{4}FET came from its four independent gates, two of which provide vertical MOS (metal-oxide-semiconductor) field-effect action whereas the other two gates provide lateral junction field effect transistor (JFET) functionality. The unique G^{4}FET structure can be leveraged to design circuits for different analog, mixed-signal and digital applications with significantly reduced transistor counts. Some of these have already been experimentally demonstrated including LC oscillators and Schmitt trigger circuit with adjustable hysteresis using negative differential resistance [5], high voltage current mirrors and differential amplifiers [6], four quadrant analog multipliers [7], adjustable threshold inverters, real time reconfigurable logic gates and DRAM cell [8], universal and programmable logic gate capable of highly efficient full adder design [9], and temperature compensated voltage references [10]. Another exciting application is the formation of quantum wire with low subthreshold swing, high mobility and low noise in depletion-all-around action when the vertical MOS gates and lateral JFET gates are used simultaneously to create a conducting channel surrounded by depletion regions [11]. G^{4}FET inspired multiple state electrostatically formed nanowires have already been used for threshold logic functions [12, 13], femtomolar bio-marker detection [14] and high-sensitivity gas sensing [15].

A G^{4}FET transistor can operate in different regimes based on the bias voltages on its four gates and the silicon epi layer thickness. A SPICE model that can work sufficiently well throughout the different operating regions and different types of simulations is absolutely necessary to design innovative circuits with G^{4}FETs. Although a number of works on G^{4}FET modeling have been reported in the literature, so far, it has not been possible to come up with a reasonably concise compact model based on device physics which is valid for different combinations of gate biases. It is theoretically possible to solve coupled non-linear differential equations describing device operation to explore different operating conditions. But this will take prohibitively long time and therefore, is not suitable for useful circuit design. However, we can find in the literature that G^{4}FET has previously been called MOSJFET [4] since it combines both metal-oxide-semiconductor field-effect transistor (MOSFET) and junction field-effect transistor (JFET) actions in a single body. This is motivation behind this proposed approach to combine existing MOSFET and JFET models for building a macromodel where the interactions between different gates are accounted for using analytical expression from device physics. This approach can provide reasonably fast and accurate DC, transient and AC circuit simulation.

The remainder of the paper is as follows — a brief summary of prior works on macromodels and G^{4}FET modeling is given in Sec. 2. Section 3 describes the device structure and operating principle of G^{4}FET. The model formulation is described in Sec. 4. Section 5 describes the results from innovative circuit simulation using G^{4}FET and validates the model by comparison with experimental results. Finally, Sec. 6 concludes the paper with a summary of the work.

2. Prior Works

2.1. *Macromodel*

The number of elements in today's integrated circuit can range from several dozens to hundreds of millions. If each individual element is modelled separately, the simulation run time will be prohibitively long. Macromodels are used to simplify circuits in a way so that the desired behavioral characteristics remain the same for all practical purposes while the computational time gets substantially reduced. Important circuit blocks such as operational amplifiers (op-amp) and comparators are usually employed in simulators using their macromodels. The need for macromodel in IC subsystem design is discussed in [16]. The authors developed a macromodel for integrated circuit op-amp with an excellent pin-for-pin representation. The model uses common elements available in most circuit simulators and it is a factor of more than six times less complex, an order of magnitude faster and less costly compared to op amp models at the electronic device level.

Logic simulation and macromodels have also been developed for digital logic blocks ([17], [18]). A behavioral multiport macromodel for the input buffers of digital integrated circuits is presented in [19] which offers comparable accuracy and improved efficiency compared to the transistor-level models. A macromodel for integrated-circuit comparators, capable of providing up to an order of magnitude reduction in CPU time and matrix size for CAD, has been reported in [20]. A lumped parameter macromodel has been derived from transistor characterization data to use in SPICE analyses for predicting the single-event upset thresholds for Texas Instruments SIMOX (Separation by IMplantation of OXygen) SOI SRAMs [21]. Physico-chemical model of the ISFET (ion-sensitive field-effect transistor) has been developed in [22] using a behavioral macromodel that can be used in commercial SPICE programs. The main goal is to get rid of the drawbacks associated with developing built-in models such as the availability of the program source, a deep knowledge of the code subroutines and structure, and the requirement of compiling the entire program for a new model implementation. An empirical macromodel for a *p*-channel floating-gate MOS synapse transistor simulation consisting of a transistor and controlled sources has been proposed in [23]. In [24], an improved SPICE macro-model for laterally diffused MOS (LDMOS) device has been proposed with better performance compared to existing BSIM3 models in both DC and AC regions. SPICE macro-modeling techniques have also been used in [25] for the compact simulation of single electronic circuits.

2.2. *G⁴FET Modeling*

Since its discovery, different approaches have been adopted to model the characteristics of G⁴FET which is briefly described in this section. The charge coupling between front, back and lateral junction-gates was considered and a 2-D analytical relationship for the fully-depleted body potential was derived in [26]. This work also derived a closed form front-interface threshold voltage expression as a function of the back and the lateral gate voltages for different back interface conditions such as accumulation, depletion and inversion. A charge sheet model has been proposed to analyze the transistor characteristics of fully-depleted G⁴FETs [27]. Here, surface accumulation behavior, drain current and gate capacitance of fully-depleted G⁴FET are modelled analytically. In [28], a mathematical model is developed to determine the subthreshold swing of thin-film fully-depleted G⁴FET. Multidimensional Lagrange and Bernstein polynomial approaches have been used in [29] for modeling and SPICE implementation. The thin film fully-depleted version of the G⁴FET has been introduced and its characteristics have been systematically investigated in [30]. The authors in [31] have reported single multivariate regression polynomial approach for modeling the device operation across different operating regimes. A physics-based simplified analytical DC model suitable for DAA (depletion-all-around) operation has been proposed in [32]. A multidimensional linear and cubic spline interpolation method for DC modeling and CAD implementation has been reported in [33].

3. Device Structure

G⁴FETs can be fabricated using standard partially or fully-depleted SOI (PD/FD-SOI) process. It has four independent gates for modulating channel conduction. There are two lateral junction-gates which act like JFET gates and two vertical oxide gates which act like MOS gates. This transistor has also been called MOSJFET [4] since it combines both metal-oxide-semiconductor field-effect transistor (MOSFET) and junction field-effect transistor (JFET) actions in a single silicon island.

G⁴FET is a majority carrier device in which a regular *n*-channel SOI MOSFET with two body contacts on the opposite sides of the channel works as a *p*-channel G⁴FET. The *n*+ doped source and drain of the MOSFET function as lateral junction-gates. They are used like JFET gates to control the channel conduction width. The top oxide gate works like a classical MOS gate whereas the buried oxide along with the substrate biasing acts as

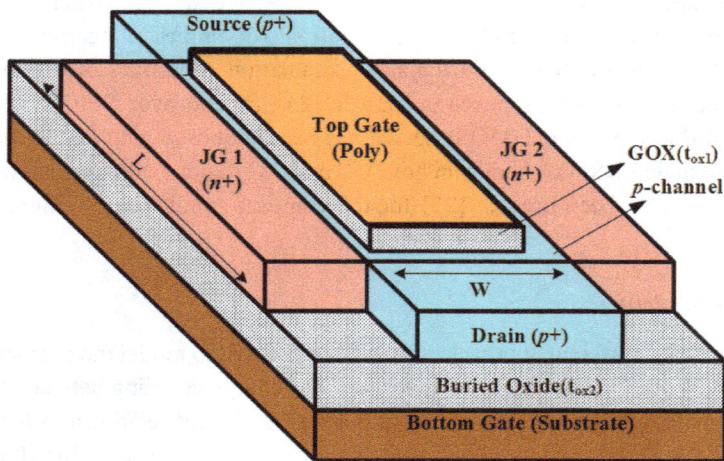

Fig. 1. Schematic Device Structure of a *p*-channel G⁴FET.

a bottom-gate. These vertical gates are used to create the accumulation/depletion/inversion of free carriers in the silicon epi layer near the top and the bottom oxide interfaces. The body contacts are highly doped to make ohmic contact with the channel and are used as the source and the drain for the *p*-channel G⁴FET. An accumulation/depletion-mode *p*-channel G⁴FET is thus realized from an inversion-mode, *n*-channel MOSFET. Similarly, an *n*-channel G⁴FET can be constructed from a conventional SOI *p*-channel MOSFET. Figure 1 shows the 3-D schematic structure of a *p*-channel G⁴FET. It is evident that no specialized fabrication procedure is necessary for this device.

4. Operating Principle and Conduction Mechanism

The device operation of G⁴FET is similar to an accumulation mode MOSFET with two junction-gates providing JFET like control on the conduction channel. The flow of the drain current is perpendicular to the conventional MOSFET current flow. The length and the width of the G⁴FET are the width and the length of original MOSFET, respectively. There can be three conduction paths, namely, 1) top surface conduction near gate oxide interface, 2) bottom surface conduction near buried oxide interface, and 3) volume conduction away from vertical oxide interfaces. Depending on various applications, the specific components can be turned on or off using appropriate gate biases.

Numerical simulation in TCAD Sentaurus is used to visually demonstrate the effects of different gate biases on the conduction path. A three-dimensional *n*-channel G⁴FET structure is created using Sentaurus Structure Editor and simulated using Sentaurus Device. The cross section halfway along the channel length is used here to demonstrate the effect of lateral and vertical biases on conduction path. The channel conduction depends on the concentration of electrons inside the channel which is shifted by different gate biases. Here, V_{TG} is the top-gate voltage, V_{BG} is the bottom-gate voltage and V_{JG} is the junction-gate voltage applied at both junction gates which are connected together. As can be seen in these figures, all the gate voltages can independently affect the conduction path and consequently, the device operation of G⁴FET.

eDensity (cm^-3)

1.073e+20
5.094e+16
2.419e+13
1.148e+10
5.452e+06
2.589e+03
1.229e+00

Fig. 2. Electron density in the top silicon film at $V_{TG} = 0$V, $V_{BG} = 0$V and $V_{JG} = 0$V.

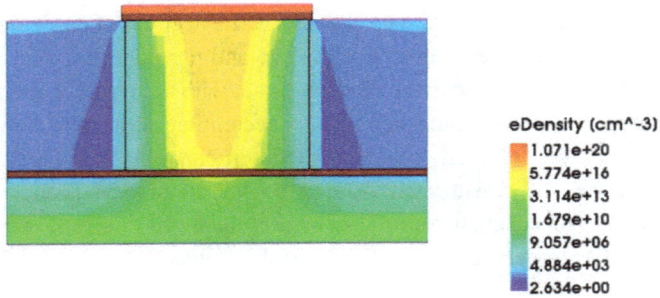

Fig. 3. Electron density in the top silicon film at V_{TG} = 0V, V_{BG} = 0V and V_{JG} = -1V.

Fig. 4. Electron density in the top silicon film at V_{TG} = 3V, V_{BG} = 0V and V_{JG} = 0V.

Fig. 5. Electron density in the top silicon film at V_{TG} = 3V, V_{BG} = 10V and V_{JG} = 0V.

Figure 2 shows the electron concentration for keeping all the biases at 0 V. Figure 3 shows the effect of lateral depletion with both junction-gate reverse biased at -1 V. The channel now becomes narrower as lateral region near junction-gates gets depleted of free carriers. Figure 4 shows the effect of accumulation caused by top-gate when V_{TG} = 3 V is applied. A thin layer of high electron density is formed near the top oxide surface. Figure 5 shows the effect of the bottom gate. Here, both the top and the bottom-gates are accumulated and the transistor provides both surface and volume conduction.

5. Model Development

G⁴FET combines MOS and JFET actions by supporting both surface and volume conduction. The top and the bottom oxide gates provide MOS action whereas the lateral junction-gates work like JFET. The threshold voltage of the top and the bottom-gates are influenced by the junction-gate voltage. It can be considered as a combination of two MOSFETs (surface conduction) working in parallel with a JFET (volume conduction).

The top-gate threshold voltage is V_{TH} and the bottom-gate voltage causing the onset of accumulation and inversion at the bottom-gate are V_{BG}^{acc} and V_{BG}^{inv}, respectively. Some of the terms used in the model are introduced below:

$$\text{Junction-gate capacitance, } C_{JG} = \varepsilon_{si}/_W$$

$$\text{Top oxide capacitance, } C_{ox1} = \varepsilon_{ox}/_{t_{ox1}}$$

$$\text{Bottom oxide capacitance, } C_{ox2} = \varepsilon_{ox}/_{t_{ox2}}$$

Three constants based on device geometry, α, β and γ are defined as,

$$\alpha = \frac{2\sqrt{2}}{\tanh(\frac{2\sqrt{2}t_{Si}}{W})}, \beta = \frac{\gamma C_{JG}/_{C_{ox1}}}{1+\alpha C_{JG}/_{C_{ox2}}}, \gamma = \frac{2\sqrt{2}}{\sinh(\frac{2\sqrt{2}t_{Si}}{W})}$$

Other terms include,

$$\varphi_F = -V_T\ln\left(\frac{N_d}{n_i}\right), \quad \varphi_b = \frac{E_g}{2} + V_T\ln\left(\frac{N_d}{n_i}\right), \quad V_P = \varphi_b - \frac{qN_dW^2}{8\varepsilon_{Si}}$$

Here, W is the width of the transistor, t_{si} is the silicon film thickness, t_{ox1} is the top oxide thickness, t_{ox2} is the buried oxide thickness, $V_T = kT/q$ is the thermal voltage, N_d is the donor concentration in the body, n_i is the intrinsic carrier concentration, ε_{si} is the permittivity of silicon, and ε_{ox} is the permittivity of silicon dioxide.

The onset voltage of accumulation and inversion for the bottom-gate, V_{BG}^{acc} and V_{BG}^{inv}, can be expressed [27] as,

$$V_{BG}^{acc} = V_{FB2} + (\gamma - \alpha)\frac{C_{JG}}{C_{ox2}}\left(V_{JG} - V_P\right) \tag{1}$$

$$V_{BG}^{inv} = V_{FB2} + \left(1 + \alpha\frac{C_{JG}}{C_{ox2}}\right)2\varphi_F - (\gamma - \alpha)\frac{C_{JG}}{C_{ox2}}(V_P) + \left(1 + \gamma C_{JG}/C_{ox2}\right)V_{JG} \tag{2}$$

The back gate may be accumulated, depleted or inverted. When the bottom-gate is in inversion i.e. $V_{BG} < V_{BG}^{inv}$,

$$V_{TH} = V_{FB1} - \gamma\left(\frac{C_{JG}}{C_{ox1}}\right)(2\varphi_F + V_P) - \alpha\left(\frac{C_{JG}}{C_{ox1}}\right)(V_{JG} - V_P) \tag{3}$$

When the bottom-gate is depleted i.e. $V_{BG}^{inv} < V_{BG} < V_{BG}^{acc}$,

$$V_{TH} = V_{FB1} - \beta(V_{BG} - V_{FB2}) + (\gamma - \alpha)\left(\frac{C_{JG}}{C_{ox1}} + \beta\frac{C_{JG}}{C_{ox1}}\right)(V_{JG} - V_P) \tag{4}$$

When the bottom-gate is in accumulation i.e. $V_{BG} > V_{BG}^{acc}$,

$$V_{TH} = V_{FB1} + (\gamma - \alpha)\left(\frac{C_{JG}}{C_{ox1}}\right)(V_{JG} - V_P) \tag{5}$$

Here, V_{FB1} and V_{FB2} are the flat band voltages of the top-gate and the bottom gates, respectively.

Based on the above relationships among different gates, a macromodel subcircuit is created combining conventional SPICE Level-1 MOSFET and the JFET models which follow from the quadratic FET model of Schichman and Hodges [34]. Of course, the model can be further improved by using higher level MOSFET models such as Level-2/3 and BSIM/BSIMSOI models for deep submicron devices. Simple models are used to show the feasibility of the macromodel demonstrating its effectiveness in capturing the essence of the complex interaction between multiple gates. Since the accumulated back gate provides a shunt leakage conduction path which is undesirable for most practical applications, it is assumed that the back gate is never accumulated and the condition for depleted or inverted back surface is considered. In the model, the top conduction is modeled using a MOSFET and the volume conduction is modeled using a JFET. However, instead of a constant threshold MOSFET, the subcircuit allows for threshold voltage modification based on multiple gate biases using the relationship described above.

6. Model Validation

The developed macromodel has been implemented in SPICE simulator as a subcircuit and validated against experimental results from five different analog and digital circuits. These include 1) negative differential resistance circuit, 2) high voltage differential amplifier, 3) Four Quadrant Analog Multiplier, 4) multiple threshold inverter, 5) G⁴FET as a universal and programmable logic gate and 6) G⁴FET full adder circuit.

6.1. *Voltage Controlled Negative Differential Resistance Circuit*

A voltage controlled negative differential resistance (VC-NDR) circuit has been simulated with the developed model. The schematic of the conventional, two-terminal NDR device, known as "lambda diode" [36], is shown in Fig. 6(a). NDR parameters such as peak/valley voltages and peak current are controlled by pinch-off voltages and transconductance of each JFET which remain fixed for a chosen pair of JFETs. The G⁴-NDR is obtained by replacing the JFETs with complementary G⁴-FETs as shown in Fig. 6(b) with the junction-gates being tied together. It has four terminals, the additional two terminals consisting of the top-gates of the n-channel and the p-channel G⁴FETs, driven by the voltages V_n and V_p, respectively. The bottom gate is not shown. A simplified schematic is shown in Fig. 6(c).

NDR parameters such as peak/valley voltages and peak current depend on the pinch-off voltage and transconductance of each JFET. In G⁴-NDR, these parameters can be modulated by the MOS front gates (V_n, V_p) resulting in a significant improvement in functionality compared to the lambda diode.

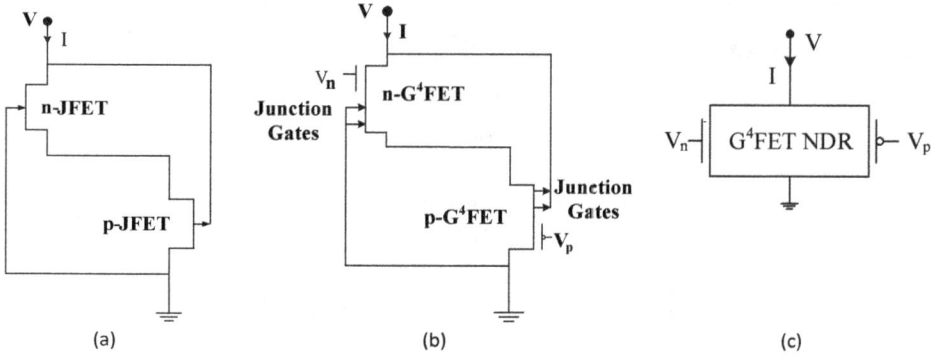

Fig. 6. Negative Differential Resistance circuit using G⁴FET, (a) A conventional two-terminal JFET NDR device, (b) a four-terminal G⁴FET NDR device, (c) a simplified symbol of G⁴FET NDR.

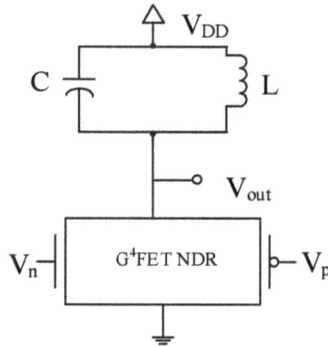

Fig. 7. An LC oscillator using G⁴-NDR.

Figure 7 shows a G⁴-NDR connected to a LC tank load which works as an LC oscillator. It was experimentally demonstrated with $V_{DD} = 3.3$ V, $L = 0.4$ mH, $C = 110$ pF [5]. The simulation output is 2.45 V_{p-p} compared with the experimental result of 2.5 V_{p-p} with a relative error of 2% as shown in Fig. 8. The extra two terminals, V_n and V_p provide additional functionality and can be used for amplitude modulation as shown in Fig. 9. Another potential application is to leverage this tunable tent map shape characteristics of the voltage controlled NDR circuit to build chaotic logic gates for hardware security applications [36].

(a)   (b)

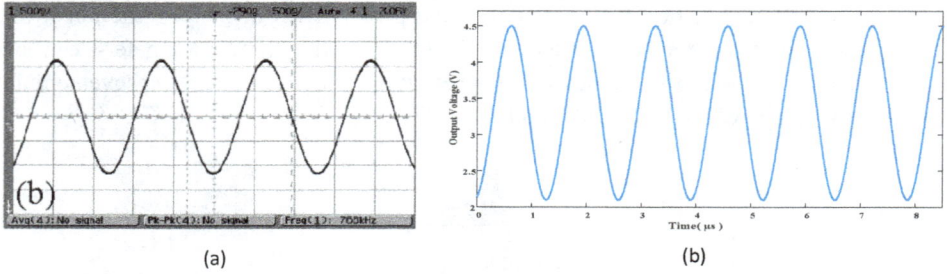

Fig. 8. Output of NDR LC oscillator; (a) experimental, (b) simulation.

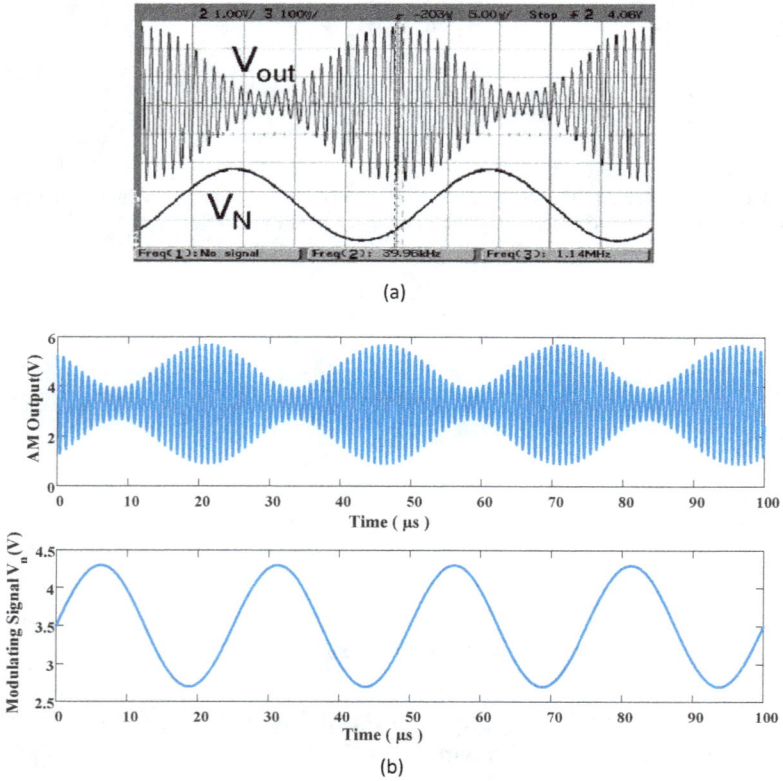

(a)

(b)

Fig. 9. Amplitude Modulated (AM) signal; (a) experimental, (b) simulation.

6.2. *Differential Amplifier*

The second circuit is a high voltage differential amplifier first demonstrated in [6]. It is simulated using a SPICE simulator. Compared to regular MOSFETS, G^{4}FETs can sustain much higher voltages when realized in the same process technology. The circuit has a current mirror for biasing which designed using two *n*-channel G^{4}FETs and a differential pair implemented using a pair of *n*-channel G^{4}FETs. A pair of *p*-channel G^{4}FETs is used as active load. The schematic is shown in Fig. 10(a). Here, the junction-gates are connected together and shown as a single gate and the unused bottom-gate is not shown for simplification.

The high voltage (HV) differential amplifier in [6] is used in a non-inverting unity gain configuration to an input of 1 V_{p-p} square wave of 1 KHz frequency. The circuit has been simulated with the developed regression model and the result in Fig. 10(b) with an output of 0.98 V_{p-p} shows good agreement with the experimental result of 1 V_{p-p} with a relative error of 2%.

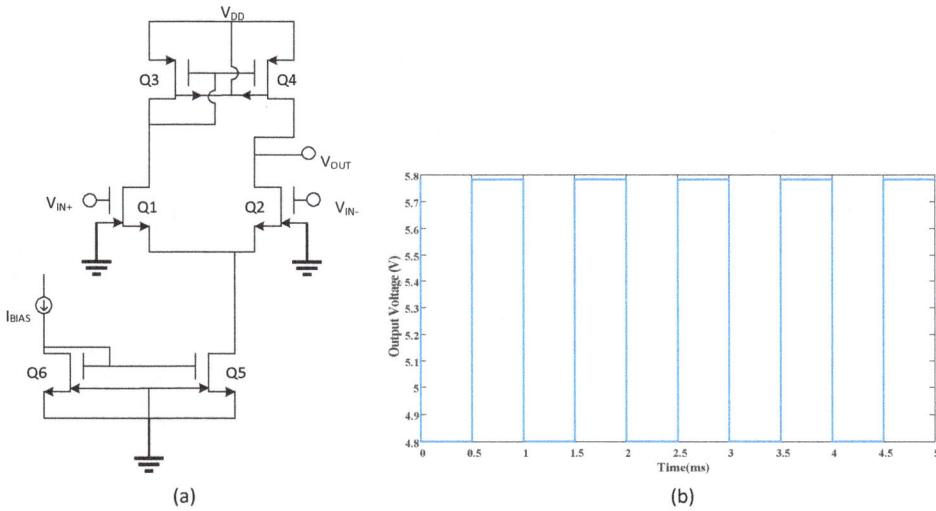

Fig. 10. G^{4}FET Differential Amplifier, a) Schematic of HV G^{4}FET differential amplifier (Q1,2: 0.3 × 10/2.4 μm, V_{JG} = 0 V, V_{BG} = 0 V; Q3,4: 0.35 μm × 2/10 μm, V_{JG} = 0 V, V_{BG} = 0 V. Q5,6: 0.3 μm × 10/2.4 μm, V_{JG} = V_{DD}, V_{BG} = 0 V), b) Output of G^{4}FET Differential amplifier (0.98 V_{p-p} compared to experimental value of 1 V_{p-p}) in non-inverting unity gain configuration (V_{DD} = 10 V, V_{in} = 1 V_{p-p} square wave with 6 V offset).

6.3. *Four Quadrant Analog Multiplier*

G^{4}FET can be used to design analog multiplier with only four transistors and two different configurations were experimentally demonstrated in [6]. The idea is to leverage the linear modulation of the front gate threshold voltage by the junction gates. In both configurations, multiplier core is made of four G^{4}FETs biased by a constant current sink and loaded by a load resistor R_L which converts the differential output current to a differential output voltage. However, the input is different for these two cases. As shown in Fig. 11(a), configuration 1 has one input V_{in1} at the top-gate and other input V_{in2} at the junction-gates, which are tied together. In configuration 2, shown in Fig. 11(b) the junction-gates are independent and two differential input voltages V_{in1} and V_{in2} are connected to two lateral junction-gates, whereas the top-gate is biased at a constant voltage.

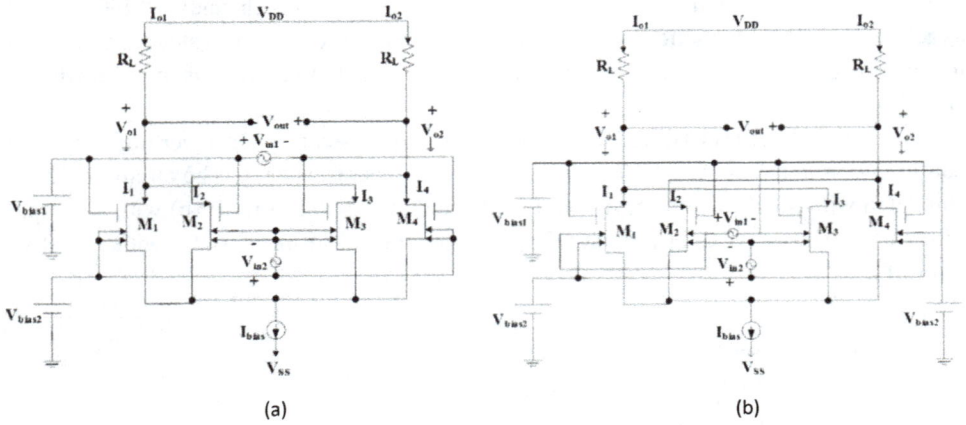

Fig. 11. Analog Multiplier with four G^{4}FETs, (a) Configuration 1, (b) Configuration 2 [33].

The macromodel was used to simulate both the configurations. DC transfer characteristic for configuration 1 and configuration 2 are shown in Fig. 12 and Fig. 13, respectively. The experimental results are shown in Fig. 12(a) and Fig. 13(a) whereas simulation results obtained using the macromodel are shown in Fig. 12(b) and Fig. 13(b), respectively. Table 1 gives a quantitative comparison between DC transfer characteristics from measurement and simulation results for both configurations. The figures and the table show that the model captures the DC transfer characteristics quite well.

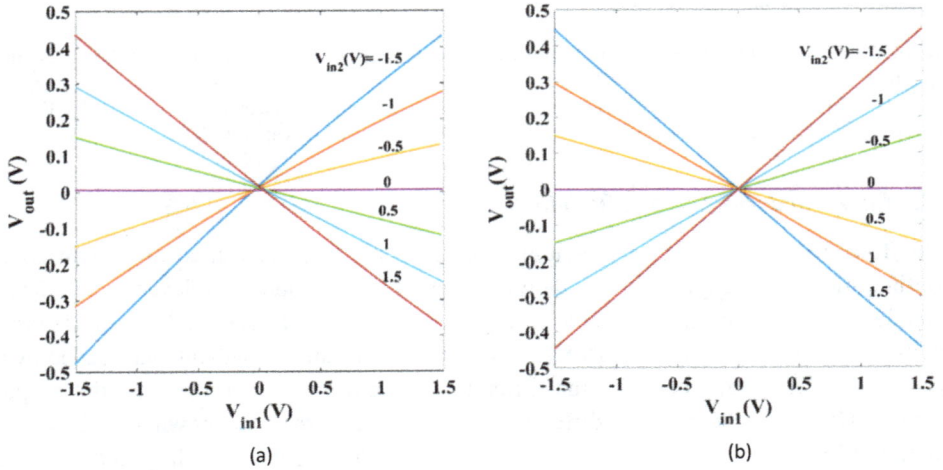

Fig. 12. DC transfer characteristic for Configuration 1; (a) Experimental, (b) Simulation.

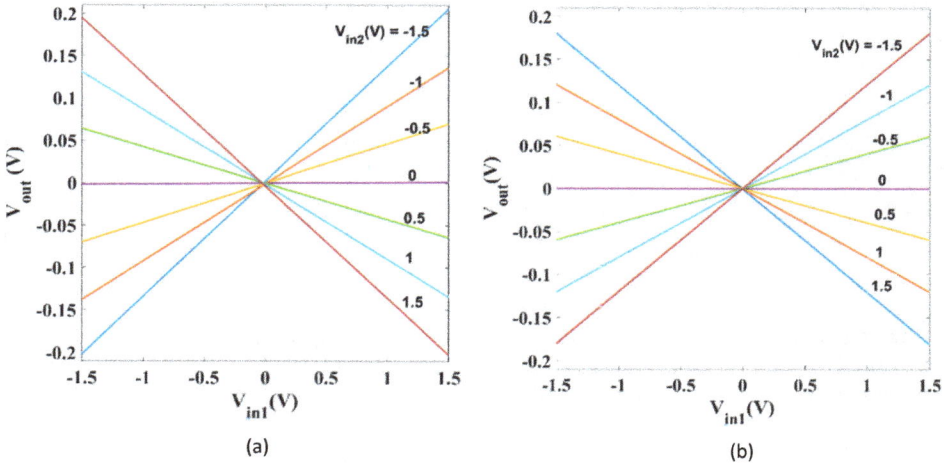

Fig. 13. DC transfer characteristic for Configuration 2; a) Experimental, b) Simulation.

Table 1. Comparison between experimental and simulation results for DC transfer characteristics.

V_{in2} (V)	Configuration 1, $V_{out\,(p\text{-}p)}$ (V)			Configuration 2, $V_{out\,(p\text{-}p)}$ (V)		
	Experiment	Simulation	Error (%)	Experiment	Simulation	Error (%)
-1.5	0.911	0.892	2.09	0.41	0.36	11.52
-1	0.594	0.594	0	0.274	0.241	12.04
-0.5	0.281	0.297	5.69	0.139	0.12	13.67
0	0	0	0	0	0	0
0.5	0.273	0.297	8.79	0.129	0.12	6.98
1	0.544	0.594	9.19	0.27	0.24	9.06
1.5	0.813	0.892	9.72	0.4	0.36	9.3

Figure 14 shows the result for configuration 1 as an analog multiplier where the inputs are a 20 Hz, 1 $V_{p\text{-}p}$ sinusoidal-wave and a 500 Hz, 1 $V_{p\text{-}p}$ square-wave. The simulation result using the macromodel in Fig. 14(b) shows good matching with the experimental result [7]. The peak-to-peak output voltage in simulation is 1.0025 V which is very close to the measurement result of 1 V, with a relative error of 0.25%.

Configuration 2 was also used as an analog multiplier with two different input signals; a 10 Hz, 4 $V_{p\text{-}p}$ triangular-wave and a 200 Hz, 4 $V_{p\text{-}p}$ square-wave. This configuration has a reduced gain compared to configuration 1, but it has a higher input voltage swing capability. Figure 15 shows the simulation results. There is a reasonable agreement in peak-to-peak output voltage between simulation results of 0.294 V and experimental measurements of 0.3 V [7] with a relative error of 2%.

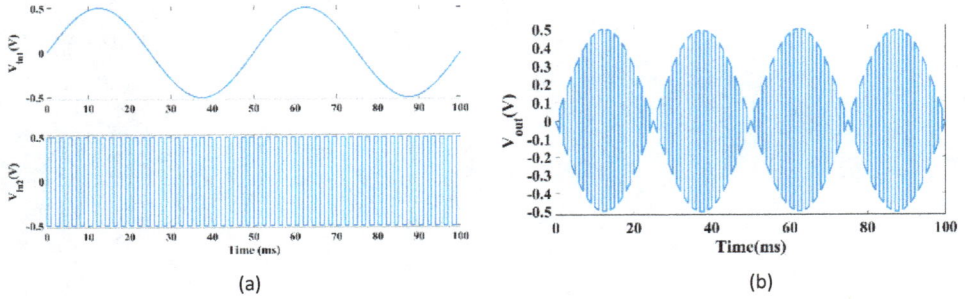

(a)   (b)

Fig. 14. Transient simulation results for product of a 20 Hz, 1 V_{p-p} sinusoidal-wave with 500 Hz, 1 V_{p-p} square-wave ($W = 0.3$ μm × 10, $L = 2.4$ μm, $V_{DD} = 3.5$ V, $V_{SS} = -3.5$ V, $I_{bias} = 35$ μA, $V_{bias1} = 2$ V, $V_{bias2} = -2.5$ V, $R_L = 100$ kΩ) for configuration 1; (a) input, (b) output.

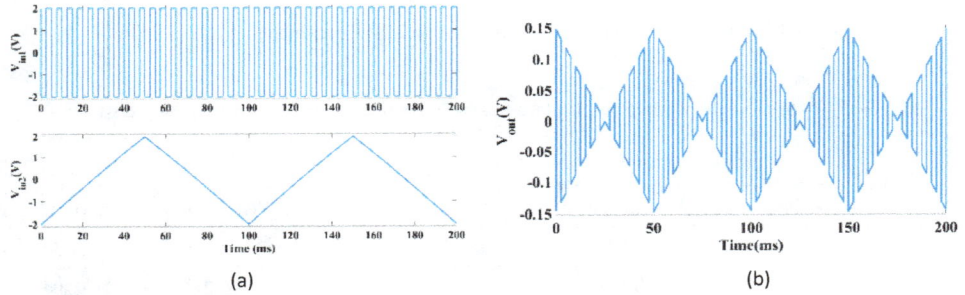

(a)   (b)

Fig. 15. Transient simulation results for product of a 10 Hz, 4 V_{p-p} triangular-wave with 200 Hz, 4 V_{p-p} square-wave ($W = 0.35$ μm, $L = 5$ μm, $V_{DD} = 5$ V, $I_{bias} = 15$ μA, $V_{bias1} = 0$ V, $V_{bias2} = -3.5$ V, $R_L = 200$ kΩ) for configuration 2; (a) input, (b) output.

6.4. *Multi-Threshold Inverter*

G^{4}FET can be used to build interesting digital circuits as well. The multiple gates offer plenty of opportunities for innovative digital designs. In [8], a multi-threshold inverter has been demonstrated. The schematic is shown in Fig. 16(a). The top-gate works as a conventional MOS gate whereas the junction-gate bias is used to change the threshold of the inverter. Three different thresholds for different junction-gate combinations are obtained. The macromodel reproduces the results in [8] quite well as shown in Fig. 16(b). Different threshold curves for varying the junction-gate voltages are shown as A, B and C. Here, V_{LG} and V_{RG} stand for left junction gate and right junction-gate bias voltages, respectively. The difference between the two successive thresholds is approximately 250 mV in this example but can modified by changing device width or oxide thickness. This can be used as a quaternary down literal converter, the basic building block of quaternary logic circuits [37].

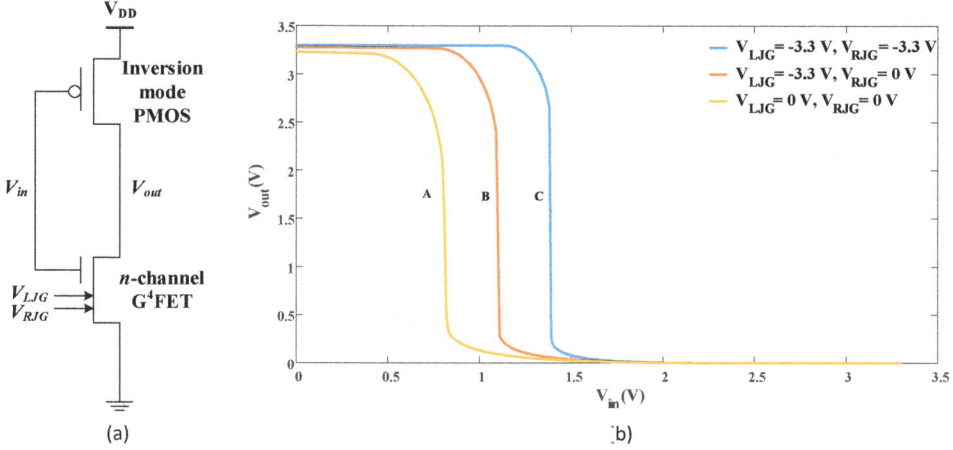

Fig. 16. (a) Schematic of a multi-threshold Inverter, (b) Output of a multi-threshold inverter.

6.5.   *Universal and Programmable Gate*

G⁴FET can be used as a universal and real-time reconfigurable logic gate as demonstrated in [8, 9]. The schematic of a programmable logic gate is shown in Fig. 17. Here, the left and the right junctions act as inputs and the top-gate acts as controller. Based on the value of the top-gate voltage, this can function as either NAND or NOR gate. Hence, in principle, G⁴FET is a universal gate, since any logic function can be computed using it.

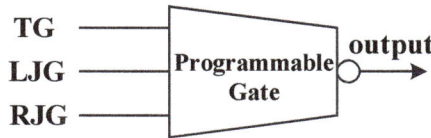

Fig. 17. Symbol of a G⁴FET programmable gate.

Figure 18 shows the results for a programmable gate using the macromodel. The output is a NAND function, $V_{out} = \overline{V_{RJG}.V_{LJG}}$, when $V_{INV,A} < V_{TG} = 0.9\ V < V_{INV,B}$. The output is a NOR function, $V_{out} = \overline{V_{RJG} + V_{LJG}}$, when $V_{INV,B} < V_{TG} = 1.2\ V < V_{INV,C}$. The results show excellent matching with the experimental results reported in [8].

6.6.   *G⁴FET Full Adder*

A full adder circuit was demonstrated in [9] using only 3 G⁴FET transistors and 2 inverters. This design drastically reduces the number of transistor count and paves the way for more compact arithmetic logic operation circuits. This reduction is enabled by unique programmable feature of G⁴FET leveraging its multi-gate functionality. Figure 19 shows the schematic of the proposed design. The circuit has been simulated using the macromodel and the output is shown in Fig. 20. The full adder functionality for all the possible combinations from the truth table in Table 2 is demonstrated and the result matches with experimental results shown in [9].

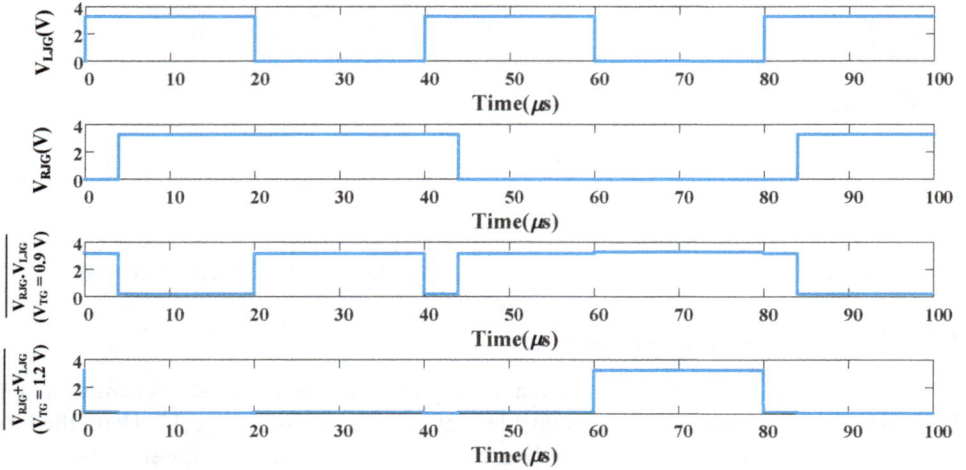

Fig. 18. Output of a programmable gate.

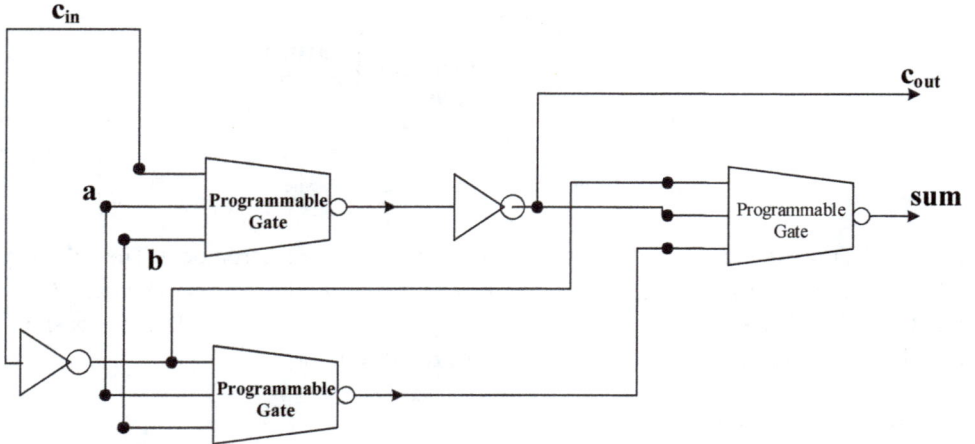

Fig. 19. Schematic of the G⁴FET full adder.

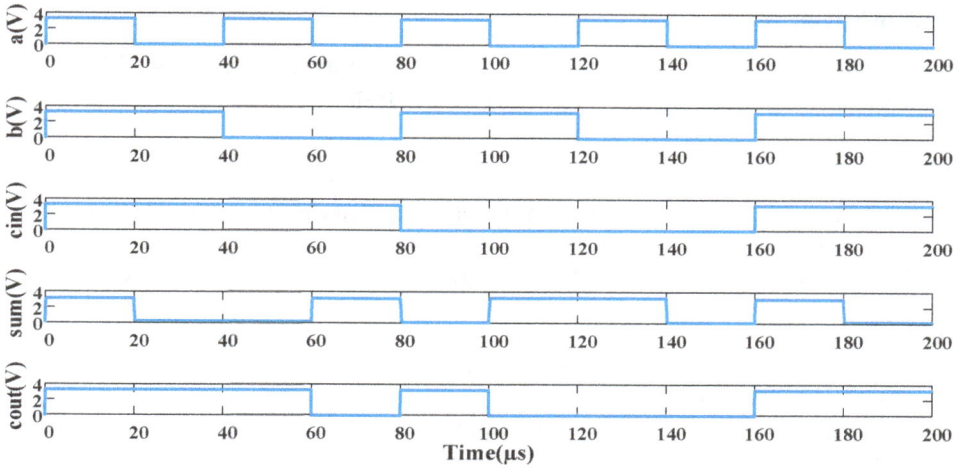

Fig. 20. Output from the full adder.

Table 2. Truth Table of a Full Adder.

a	b	c_in	sum	c_out
0	0	0	0	0
0	0	1	1	0
0	1	0	1	0
0	1	1	0	1
1	0	0	1	0
1	0	1	0	1
1	1	0	0	1
1	1	1	1	1

7. Conclusion

The macromodel of G⁴FET takes the underlying physics of G⁴FET operation into account and effectively emulates the interaction between multiple independent gates by combining MOSFET and JFET models into a single subcircuit. The existence of robust, stable and accurate MOSFET and JFET SPICE models facilitates faster implementation in circuit simulator. As shown in this work, the model has successfully reproduced experimental findings from a number of analog and digital G⁴FET circuits. Since this model captures the essential interplay between multiple gates, it has the potential to enable circuit designers to come up with new interesting analog and digital circuits using G⁴FETs. In this work, the macromodel has been developed with Level 1 SPICE models to show the feasibility of such an approach. Further improvements in accuracy can be accomplished using higher level models with additional optimized fitting parameters.

Acknowledgments

The authors would like to thank Dr. Benjamin Blalock from the University of Tennessee for helpful discussions and suggestions on G⁴FET modeling.

References

1. G. E. Moore, "Cramming more components onto integrated circuits", *Electronics*, vol. 38, no. 8, pp. 114-117, April 19, 1965.
2. J.-P. Colinge, Silicon-on-Insulator Technology: Materials to VLSI, 2nd ed., Norwell, MA, USA: Kluwer, 1997.
3. S. Cristoloveanu, B. Blalock, F. Allibert, B. Dufrene and M. Mojarradi, "The Four-Gate Transistor", in *Proceedings of the 2002 European Solid-State Device Research Conference*, pp. 323-326, Firenze, Italy, September 2002.
4. B. J. Blalock, S. Cristoloveanu, B. M. Dufrene, F. Allibert, and M. M. Mojarradi, "The Multiple-Gate MOS-JFET Transistor", *International Journal of High Speed Electronics and Systems*, vol. 12, no. 2, pp. 511-520, 2002.
5. K. Akarvardar, S. Chen, J. Vandersand, B. J. Blalock, R. D. Schrimpf, B. Prothro, C. Britton, S. Cristoloveanu, P. Gentil, and M. M. Mojarradi, "Four-gate transistor voltage-controlled negative differential resistance device and related circuit applications", in *Proc. IEEE International SOI Conference*, pp. 71-72, 2006.
6. S. Chen, J. Vandersand, B. J. Blalock, K. Akarvardar, S. Cristoloveanu and M. Mojarradi, "SOI Four-Gate Transistors (G4-FETs) for High Voltage Analog Applications", *Proceedings of ESSCIRC*, Grenoble, France, 2005.
7. K. Akarvardar, S. Chen, B.J. Blalock, S. Cristoloveanu, P. Gentil, M. Mojarradi, "A Novel Four-Quadrant Analog Multiplier Using SOI Four-Gate Transistors (G4-FETs)", *Proceedings of ESSCIRC*, Grenoble, France, 2005.
8. K. Akarvardar, B. Blalock, S. Chen, S. Cristoloveanu, P. Gentil and M. M. Mojarradi, "Digital Circuits Using SOI Four-Gate Transistor", *2006 8th International Conference on Solid-State and Integrated Circuit Technology Proceedings*, Shanghai, 2006, pp. 1867-1869.
9. A. Fijany, F. Vatan, M. Mojarradi, B. Toomarian, B. Blalock, K. Akarvardar, S.Cristoloveanu, and P. Gentil, "The G⁴-FET: A Universal and Programmable Logic Gate", *Proceedings of the 15th ACM Great Lakes symposium on VLSI*, pp. 349-352, 2005.
10. S. C. Terry, S. Chen, B. J. Blalock, J. R. Jackson, B. M. Dufrene, M. M. Mojarradi, S. K. Islam and M. N. Ericson, "Temperature-Compensated Reference Circuits for SOI", *2004 IEEE International SOI Conference*, October 2004.
11. K. Akarvardar, S. Cristoloveanu, P. Gentil, B. J. Blalock, B. Dufrene and M. M. Mojarradi, "Depletion-all-around in SOI G/sup 4/-FETs: a conduction mechanism with high performance", *Proceedings of the 30th European Solid-State Circuits Conference*, Leuven, Belgium, 2004, pp. 217-220.
12. G. Segev, I. Amit, A. Godkin, A. Henning and Y. Rosenwaks, "Multiple State Electrostatically Formed Nanowire Transistors", in *IEEE Electron Device Letters*, vol. 36, no. 7, pp. 651-653, July 2015.
13. J. S. Friedman, A. Godkin, A. Henning, Y. Vaknin, Y. Rosenwaks and A. V. Sahakian, "Threshold Logic With Electrostatically Formed Nanowires", in *IEEE Transactions on Electron Devices*, vol. 63, no. 3, pp. 1388-1391, March 2016.

14. G. Shalev, G. Landman, I. Amit, Y. Rosenwaks, and I. Levy, 'Specific and label-free femtomolar biomarker detection with an electrostatically formed nanowire biosensor', *NPG Asia Mater.*, 5(3), p. e41, 2013.

15. A. Henning, N. Swaminathan, N., Godkin, G. Shalev, I. Amit, and Y. Rosenwaks, 'Tunable diameter electrostatically formed nanowire for high sensitivity gas sensing', *Nano Research*, 8(7), pp. 2206-2215, 2015.

16. G. R. Boyle, D. O. Pederson, B. M. Cohn, and J. E. Solomon, "Macromodeling of integrated circuit operational amplifiers", *IEEE Journal of Solid-State Circuits*, vol. 9, no. 6, pp. 353-364, Dec. 1974.

17. J. R. Greenbaum, "Digital IC models for computer-aided design", *Electronics*, vol. 46, no. 25 (1973): 121-125.

18. D. N. Pocock and M. G. Krebs, "Terminal Modeling and Photocompensation of Complex Microcircuits", *IEEE Transactions on Nuclear Science*, vol. 19, no. 6, pp. 86-93, Dec. 1972.

19. I. S. Stievano, F. G. Canavero, and I. A. Maio, "Behavioural macromodels of digital IC receivers for analogue-mixed signal simulations", in *Electronics Letters*, vol. 41, no. 7, pp. 396-397, 31 March 2005.

20. I. E. Getreu, A. D. Hadiwidjaja and J. M. Brinch, "An integrated-circuit comparator macromodel", in *IEEE Journal of Solid-State Circuits*, vol. 11, no. 6, pp. 826-833, Dec. 1976.

21. S. E. Kerns et al., "Model for CMOS/SOI single-event vulnerability", in *IEEE Transactions on Nuclear Science*, vol. 36, no. 6, pp. 2305-2310, Dec. 1989.

22. S. Martinoia and G. Massobrio, "A behavioral macromodel of the ISFET in SPICE", *Sensors and Actuators B: Chemical*, vol. 62, no. 3, pp. 182-189, 2000.

23. K. Rahimi, C. Diorio, C. Hernandez, and M. D. Brockhausen, "A simulation model for floating-gate MOS synapse transistors", *2002 IEEE International Symposium on Circuits and Systems. Proceedings* (Cat. No.02CH37353), Phoenix-Scottsdale, AZ, 2002, pp. II-532-II-535 vol. 2.

24. S. F. Frere, P. Moens, B. Desoete, D. Wojciechowski, and A. J. Walton, "An improved LDMOS transistor model that accurately predicts capacitance for all bias conditions", in *Proceedings of the 2005 International Conference on Microelectronic Test Structures*, 2005, pp. 75-79.

25. Y.S. Yu, H. S. Lee, and S. W. Hwang. "SPICE macro-modeling for the compact simulation of single electron circuits", Journal of Korean Physical Society, vol. 33, pp. S269-S272, 1998.

26. K. Akarvardar, S. Cristoloveanu, and P. Gentil, "Analytical modeling of the two-dimensional potential distribution and threshold voltage of the SOI four-gate transistor", *IEEE Transaction on Electron Devices*, vol. 53, no. 10, pp. 2569-2577, October 2006.

27. S. Sayed and M. Z. R. Khan, "Analytical modeling of surface accumulation behavior of fully depleted SOI four gate transistors (G 4-FETs)", *Solid-State Electronics*, vol. 81, pp. 105-112, 2013.

28. S. Sayed, M. I. Hossain and M. Z. R. Khan, "A Subthreshold Swing Model for Thin-Film Fully Depleted SOI Four-Gate Transistors", *IEEE Transactions on Electron Devices*, vol. 59, no. 3, pp. 854-857, March 2012.

29. M. S. Hasan, T. Rahman, S. K. Islam, and B. J. Blalock, "Numerical modeling and implementation in circuit simulator of SOI four-gate transistor (G4FET) using multidimensional Lagrange and Bernstein polynomial", *Microelectronics Journal*, vol. 65, pp. 84-93, July 2017.

30. K. Akarvardar, S. Cristoloveanu, M. Bawedin, P. Gentil, B. J. Blalock, and D. Flandre, "Thin film fully-depleted SOI four-gate transistors", *Solid-State Electronics*, vol. 51, no. 2, pp. 278-284, February 2007.

31. M. S. Hasan, S. K. Islam, "DC Modeling of SOI Four-Gate Transistor (G4FET) for Implementation in Circuit Simulator Using Multivariate Regression Polynomial", *IET Circuits, Devices & Systems*, Apr. 2018.

32. K. Akarvardar, S. Cristoloveanu, P. Gentil, R. D. Schrimpf and B. J. Blalock, "Depletion-All-Around Operation of the SOI Four-Gate Transistor", in *IEEE Transactions on Electron Devices*, vol. 54, no. 2, pp. 323-331, Feb. 2007.

33. M. S. Hasan, S. K. Islam and B. J. Blalock, "Modeling of SOI four-gate transistor (G 4 FET) using multidimensional spline interpolation method", *Microelectronics Journal*, vol. 76, pp. 33-42, Jun 2018.

34. H. Shichman and D. A. Hodges, "Modeling and simulation of insulated-gate field-effect transistor switching circuits", in *IEEE Journal of Solid-State Circuits*, vol. 3, no. 3, pp. 285-289, Sept. 1968.

35. H. Takagi and G. Kano, "Complementary JFET negative-resistance devices", *IEEE Journal of Solid-State Circuits*, vol. 10, no. 6, pp. 509-515, 1975.

36. M. B. Majumder, M. S. Hasan, M. Uddin, and G. S. Rose, "Chaos computing for mitigating side channel attack", *2018 IEEE International Symposium on Hardware Oriented Security and Trust (HOST)*, Washington, DC, 2018, pp. 143-146.

37. J. Shen, K. Tanno, and O. Ishizuka, "Down literal circuit with neuron-MOS transistors and its applications", *Proceedings of 29th IEEE International Symposium on Multiple-Valued Logic*, Freiburg, Germany, 1999, pp. 180-185.

Multivariate Regression Polynomial: A Versatile and Efficient Method for DC Modeling of Different Transistors (MOSFET, MESFET, HBT, HEMT and G⁴FET)

Md Sakib Hasan[*], Samira Shamsir, Mst Shamim Ara Shawkat, Frances Garcia and Syed K. Islam

Department of Electrical Engineering and Computer Science,
The University of Tennessee, Knoxville, TN 37996, USA
[*]mhasan4@utk.edu

This work presents multivariate regression polynomial as a versatile and efficient method for DC modeling of modern transistors with very different underlying physics including MOSFET (metal-oxide-semiconductor field-effect transistor), MESFET (metal–semiconductor field-effect transistor), HBT (heterojunction bipolar transistor), HEMT (High-electron-mobility transistor) and a novel silicon-on-insulator four-gate transistors (G⁴FET). A set of available data from analytic solution, TCAD simulation, and experimental measurements for different operating conditions is used to empirically determine the parameters of this model and a different set of test data is used to verify its predictive accuracy. The developed model expresses the drain current as a single multivariate regression polynomial with its validity spanning across different possible operating regions as long as the chosen independent variables lie within the range of training data set. The continuity of the resulting polynomial and its first and second order derivatives make it particularly suitable for implementation in a circuit simulator. The model also provides a method for further simplification based on prior knowledge of the underlying physical mechanism and shows excellent predictive capability for different kinds of devices. This can be very useful for modeling deep-submicron emerging devices for which any closed-form analytical solution is not yet available.

Keywords: Multivariate regression; MOSFET; MESFET; multiple-gate transistor; HBT; HEMT; semiconductor device models.

1. Introduction

Numerical models represent an alternative to physics-based analytical models for simple, fast and accurate device modeling. The methodology comprises taking measured or simulated data as input, development of model based on these empirical data and reasonably accurate reproduction of the complex nonlinear behavior of the semiconductor devices. In most cases, they are equally applicable to different types of transistors with very different operating principles and fabricated using various process technologies. There is an abundance of emerging deep-submicron devices with extremely complex underlying mechanism. It is quite difficult and time consuming to develop closed-form analytical solutions suitable for SPICE implementation for these devices. A completely physics-based model usually involves solving coupled nonlinear differential equations and for modern day circuits containing a significant number of transistors, this quickly becomes impractical. In these cases, numerical modeling can provide an alternative for faster circuit simulation.

A plethora of numerical modeling approaches have been reported in literature. The model developers have often used quadratic or higher order polynomials in their model formulation since their unique properties render them suitable for implementation in circuit simulator. This work reports a multivariate regression polynomial approach to model DC I-V characteristics of several different transistors with widely different working mechanisms. The modelled devices include long-channel MOSFETs following Level-1 SPICE model [1], deep submicron NMOS and PMOS for different channel lengths (22, 32 and 45 nm), metal–semiconductor field-effect transistors (MESFET) [2], heterojunction bipolar transistors (HBT) [3, 4], high-electron-mobility transistors (HEMT) and a novel silicon-on-insulator four-gate transistors (G^4FET) [5, 6]. First, a set of training data, obtained from analytic expression/TCAD Sentaurus simulation/experimental measurements, has been used to determine the parameters of the model. Then the model accuracy has been verified against a different set of test data. The model expresses the output current as a single multivariate polynomial where the independent variables can be terminal voltages, terminal currents (base current for BJT), geometric dimensions (W, L, t_{ox}) etc. As a polynomial, it has zeroth (C^0), first (C^1) and second (C^2) order continuity which makes the model suitable for most circuit simulators employing Newton-Raphson method for solving circuit equations. The excellent predictive capability of the model for different kinds of devices bears testimony to its versatility and effectiveness. To demonstrate this approach, we have implemented it without any prior assumption of knowledge about underlying physical mechanisms. However, even in those cases, where it is very difficult to obtain a physics based simplified final expression, it is likely to have some prior insight from device physics. Since the method allows for choosing arbitrary model terms, any prior knowledge can be leveraged to get rid of unnecessary terms to further simplify the regression polynomial. This approach can aid model developers to develop simple, fast and accurate SPICE models for deep submicron emerging devices with no closed-form analytical solution yet available which, in turn, will enable circuit designers to explore innovative applications using those devices.

2. Background

The most commonly used numerical methods utilize lookup tables and quadratic or higher order polynomials for interpolation between data points. A table lookup based approach for the empirical modeling of FETs in circuit simulators has been used in [7] to address the specific requirements of analog circuit design, such as accuracy in reproducing small-signal parameters, large signal nonlinearities, subthreshold characteristics, substrate effects, short-channel effects, and voltage dependent capacitances. In another work, a table lookup model for MOSFETs consisting of a main table, a coarse 3-D sub-table to incorporate substrate effects and a table to interpolate between channel lengths has been implemented in SPICE to overcome the inadequacies of analytical models in representing short channel effects [8].

An approach to dynamic MOSFET modeling, which is especially suited for the simulation of low-voltage mixed signal circuits has been reported [9]. The model is based on the interpolation of terminal charges and conductive currents with physically motivated functions such as piecewise polynomial and/or exponential splines based on transient current-voltage data obtained through measurement or simulation of the devices. A general *n*-dimensional first order continuous table model has been proposed in [10] where each table model has been shown to reproduce the exact behavior of the DC current expressions of two basic physical device models; the Ebers-Moll bipolar transistor model and the GLASMOST MOSFET model with high accuracy and less evaluation times. Simple interpolation methods have been developed in [11] to construct any current table from a small basis set of tables for variation of width, length, and temperature. Quadratic B-splines with not-a-knot boundary conditions were used for length and temperature interpolation, whereas simple scaling along with decomposition of channel has been done for width variation to take narrow width non-idealities into consideration.

A blending function combining exponential and polynomial interpolation for the accurate evaluation of the MOSFET drain current in the transition region between weak and strong inversions has been implemented in [12]. This model offers several interpolation methods in the table model providing the model user with a flexibility to choose based on the required simulation speed, memory consumption, and accuracy and showed good results in DC, transient, and AC analyses.

In [13], Bernstein approximation technique has been extended to multidimensional variation diminishing interpolation and applied to DC current and intrinsic charge modeling of the MOSFET to increase simulation efficiency. Lagrange and Bernstein interpolation polynomials have also been used to model multi-gate transistors [14]. Monotonic Piecewise Cubic Interpolation has been used in [15] to determine the MOSFET operating point using stored table value generated by a 2-D device simulator. In [16], the quadratic fits were used to model triode region whereas linear fits were used for the saturation region. A table-style spline formulation has been presented in [17] using quadratic splines ensuring continuity of the function and its derivative and presented a new data-compression scheme for polynomial spline coefficient storage. Multi-variate linear and cubic spline interpolation have also been used for modeling multi-gate SOI devices [18]. In another work on table method [19], a methodology of generating compact and accurate first order table model for highly nonlinear multidimensional behavior has been demonstrated. In this work, a multi-variate regression polynomial model has been presented which is capable of modeling the DC characteristics of different types of transistors with high accuracy and simplicity in developing and implementing in SPICE simulator. This technique has been previously used for modeling silicon-on-insulator multi gate transistors known as G^{4}FETs [20].

3. Model Development

In this work, multivariate regression polynomial has been used to derive the numerical model from the available data set henceforth termed as the training set. First a polynomial regression model has to be chosen with suitable model terms i.e. orders of the different independent variables. Then the coefficients of this regression model are estimated to minimize the least square error based on training data which can be obtained from experiment, TCAD simulation or analytical expression. The total number of terms in the model is chosen by user rather than predetermined by the number of sample points as done in conventional interpolation methods. This model has the flexibility to allow non-integer exponents. Moreover, prior knowledge of the expected behavior of the devices either based on device physics or experimental measurements can be leveraged to simplify model terms in the final expression.

Once the multivariate polynomial model has been specified, the problem boils down to the estimation of the vector α for the linear system of equations as follows,

$$X\alpha = y \tag{1}$$

where X is the matrix of model terms of independent variables evaluated at chosen data points, α is the vector of unknown parameters and y is the vector of known dependent variable (drain/collector current in this work).

For example, if there are two independent variables x_1 and x_2, a dependent variable y, and the dependence of y on x_1, x_2 is intended to be modelled up to second order, the model has six terms x_1, x_2, x_1^2, x_2^2, x_1x_2 and a constant term denoted by c. Each model term has a co-efficient, which is a parameter of the model, and there are six such parameters denoted by α_1, α_2... up to α_6. In addition, this can be simplified even further by ignoring the quadratic term for x_2. However, if all the model terms are chosen and a set of m observations is collected as the training set, then the equation (1) becomes-

$$\begin{bmatrix} x_1^1 & x_2^1 & (x_1^1)^2 & (x_2^1)^2 & x_1^1 x_2^1 & c \\ x_1^2 & x_2^2 & (x_1^2)^2 & (x_2^2)^2 & x_1^2 x_2^2 & c \\ \dots & \dots & \dots & \dots & \dots & \dots \\ \dots & \dots & \dots & \dots & \dots & \dots \\ x_1^m & x_2^m & (x_1^m)^2 & (x_2^m)^2 & x_1^m x_2^m & c \end{bmatrix} \begin{bmatrix} \alpha_1 \\ \alpha_2 \\ \dots \\ \dots \\ \alpha_6 \end{bmatrix} = \begin{bmatrix} y^1 \\ y^2 \\ \dots \\ \dots \\ y^m \end{bmatrix} \tag{2}$$

Here, each row in X and y corresponds to a separate observation and it is denoted using superscript. From linear algebra, unique solution for parameters is possible when the matrix X is nonsingular and the number of rows is not less than columns i.e. $m \geq 6$. Now if X is a $m \times n$ matrix, with $m \geq n$, parameters can be estimated using different techniques. Pivoted QR decomposition is used in this work, which is quite efficient, numerically stable, and provides variance for the parameters. First, the matrix X is scaled so that all columns have unit mean. We have used the diagonal scaling matrix S = diag (1/mean(X, 1)). Now equation (1) becomes,

$$XSS^{-1} \alpha = y \tag{3}$$

If a scaled parameter vector $\beta = S^{-1} \alpha$ and a scaled matrix C = XS are considered, this becomes,

$$C \beta = y \tag{4}$$

After that β can be solved and scaled back to find out α. The above-mentioned procedure is implemented in MATLAB using polyfitn toolbox [21] for generating the results shown in the following sections. The modeling order for different independent variables can be chosen by model developer, irrespective of the amount of data points taken, corresponding to the different value of those variables. If one chooses to include n independent variables $x_1, x_2, \ldots, x_n$ and highest order of an variable x_i is denoted by O_{x_i}, then maximum number of terms in the final equations can be written as,

$$N_{terms} = \prod_{i=1}^{n}(O_{x_i} + 1) \tag{5}$$

The number of required additions and multiplications for evaluating the model polynomial for a particular set of model terms will determine the speed of the circuit simulation. If the total number of additive terms is N_{add} and the total number of multiplications is N_{mul} then,

$$N_{add} = N_{terms} - 1 \tag{6}$$

$$N_{mul} = N_{terms} \times (\frac{\sum_{i=1}^{n} O_{x_i}}{2} - 1) + 1 \tag{7}$$

As a result of equations (6) and (7), the complexity of the model will increase with an increase in the order and cause a reduction in simulation speed. However, with prior knowledge of the device characteristics, the model developer may select a few model terms of suitable orders for capturing essential behavior without choosing all possible combinations of model terms up to a given order.

4. Model Validation

DC I-V characteristics of five different transistors with very different underlying physics including MOSFET, MESFET, HBT, HEMT and a novel silicon-on-insulator four-gate transistors (G^{4}FET) have been modeled using multivariate regression polynomial. Analytical model, TCAD simulation and experimental results have been used for the development and validation of the model.

4.1. *MOSFET*

Both long-channel and short-channel MOSFETs have been modeled using regression polynomial. The results are described in the following subsections.

4.1.1. *Long-channel MOSFET*

For long–channel *n*-MOSFET, Level 1 SPICE model based on Shichman and Hodges [22] equation is used to generate the data. From standard analytical expression of long channel MOSFET, it is clear that the drain current has a linear relation to device aspect ratio (*W/L*). Therefore, the drain current can be normalized to *W/L*. The normalized drain current I'_{DS} can be written as:

$$I'_{DS} = \frac{I_{DS}}{W/L} = \begin{cases} 0 \; ; V_{GS} \leq V_{TH} \\ \mu_n C_{ox} \left[(V_{GS} - V_{TH})V_{DS} - \frac{V_{DS}^2}{2} \right] (1 + \lambda V_{DS}); V_{GS} - V_{TH} \geq V_{DS} \\ \frac{\mu_n C_{ox}}{2} (V_{GS} - V_{TH})^2 (1 + \lambda V_{DS}) \; ; V_{GS} - V_{TH} \leq V_{DS} \end{cases} \tag{8}$$

Here, V_{DS} = drain-source voltage, V_{GS} = gate-source voltage, V_{TH} = threshold voltage, μ_n = electron mobility, C_{ox} = oxide capacitance/area = ε_{ox}/t_{ox}, λ = channel-length modulation parameter. The values of the parameters are: μ_n = 1000 cm^2/(V.s), C_{ox} = 3.36µF/cm^2, V_{TH} = 0.7 V, λ = 0.03 V^{-1}.

Figure 1 shows the surface plot of I'_{DS} as a function of V_{DS} and V_{GS}. The dots show the data points evaluated from equation (8). The objective is to develop a model from these limited number of data points that will predict I'_D for different values of V_{DS} and V_{GS} not present in Fig. 1. First, regression coefficients are calculated from these data points using the method described in section 3. Then drain current, I'_D for unknown drain and gate

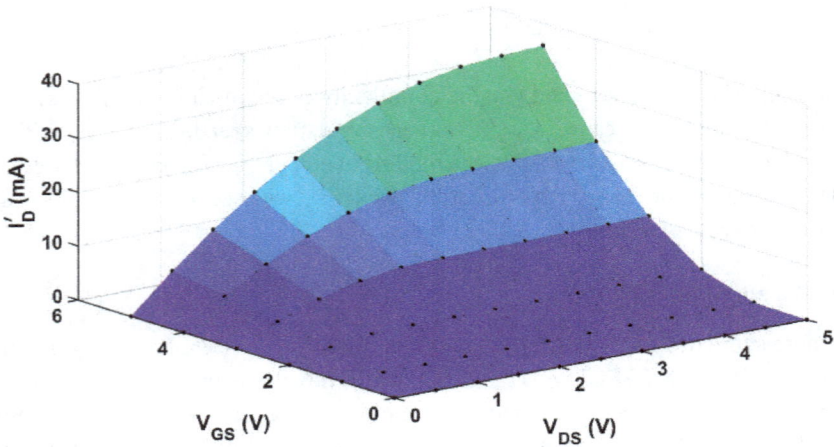

Fig. 1. Training data set for model development (long-channel *n*-MOSFET).

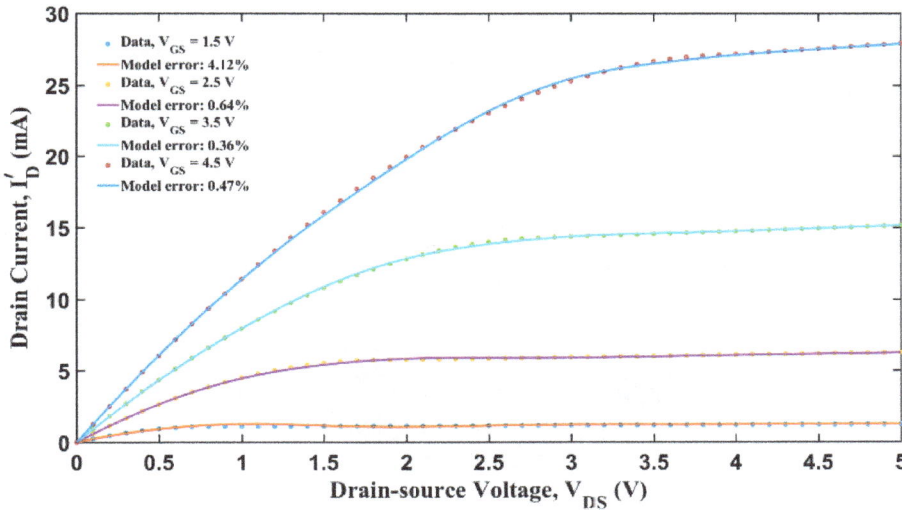

Fig. 2. Comparison between model and data for long-channel *n*-MOSFET.

biasing condition are calculated using the regression polynomial. Figure 2 shows the comparison between the results obtained from the developed model to those generated from equation (8). It shows that the values obtained from the numerical model match quite well with those obtained from equation (8).

4.1.2. *Short-channel (deep sub-micron) MOSFET*

To validate the performance of the model for short-channel MOSFETs, three sets of data for three very short channel lengths 22 nm, 32 nm and 45 nm have been generated using TCAD Sentaurus. The substrate background doping level and the peak concentration of the halo implant have been defined as $5 \times 10^{15}/cm^3$ and $4 \times 10^{18}/cm^3$, respectively. The junction depth for the halo, the extension and the source/drain implants are set as 0.05 μm, 0.008 μm and 0.04 μm, respectively. The nitride spacer length, the gate oxide thickness and the high-κ oxide thickness have been defined as 0.014 μm, 0.0006 μm and 0.0020 μm, respectively. All I–V simulations are performed using hydrodynamic transport model, where the carrier temperature equation for both electrons and holes is solved using electrostatic Poisson equation and carrier continuity equations. For the I_D–V_{DS} simulations, lattice self-heating effects are included by also solving the lattice temperature equation.

4.1.2.1. 22 nm Channel Length (NMOS)

Figure 3 shows the surface plot of I'_{DS} as a function of V_{DS} and V_{GS}. The model parameters are extracted based on these training data following the procedure outlined in Sec. 3. Then drain currents, I'_{DS} for unknown drain and gate biasing condition are calculated using the multivariate regression polynomial for two variables V_{DS} and V_{GS}. Figure 4 shows the comparison between the results obtained from the developed model to a different set of test data obtained from Sentaurus. It shows reasonably good agreement between the data and the model prediction.

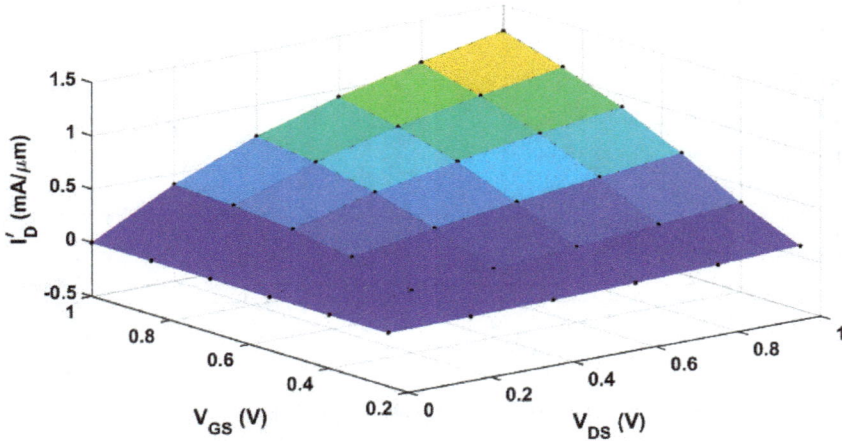

Fig. 3. Training data set for model development (22 nm *n*-MOSFET).

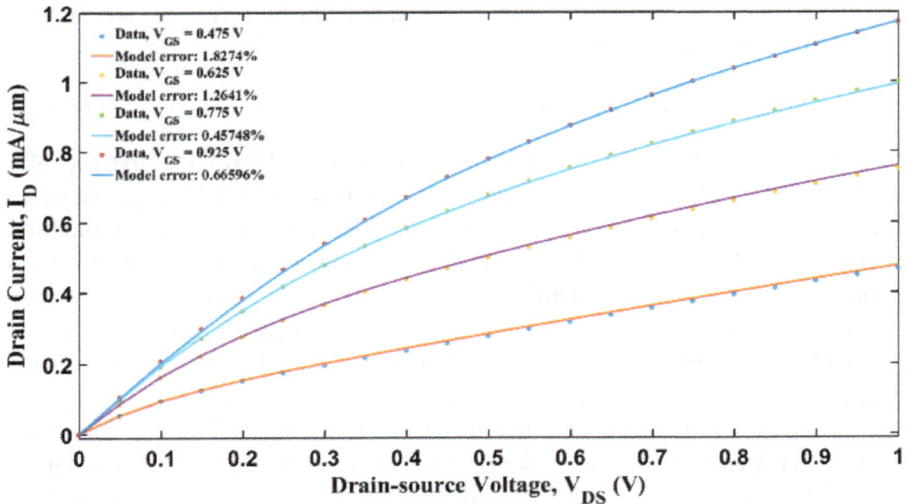

Fig. 4. Comparison between model and data for 22 nm *n*-MOSFET.

4.1.2.2. 22 nm Channel Length (PMOS)

The data for PMOS transistor has also been generated with the gate length of 22 nm by the method described in the previous section. The training data set is shown in Fig. 5 and the model predictions for a testing data set are shown in the isolines for different gate biases in Fig. 6. The relatively small mean error for different bias conditions shows good agreement between the model and the data.

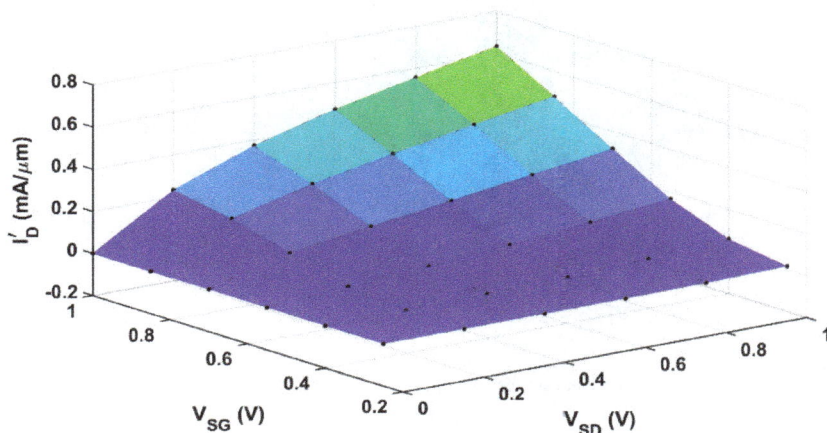

Fig. 5. Training data set for model development (22 nm p-MOSFET).

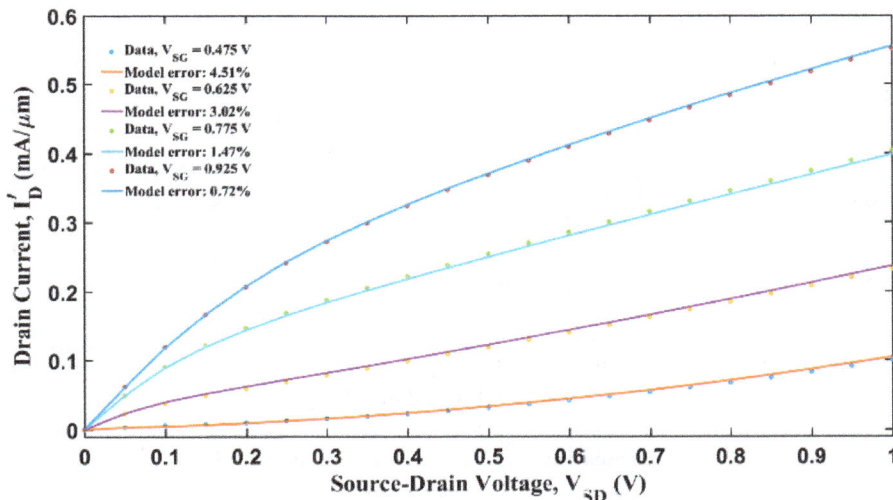

Fig. 6. Comparison between model and data for 22 nm p-MOSFET.

4.1.2.3. 32 nm Channel Length

The same measurements have also been done for 32 nm NMOS in TCAD Sentaurus. The training data set is shown in Fig. 7 and the test results for different set of gate voltages is shown Fig. 8. The relatively small value of corresponding mean model error shows good matching between the test data and the model prediction.

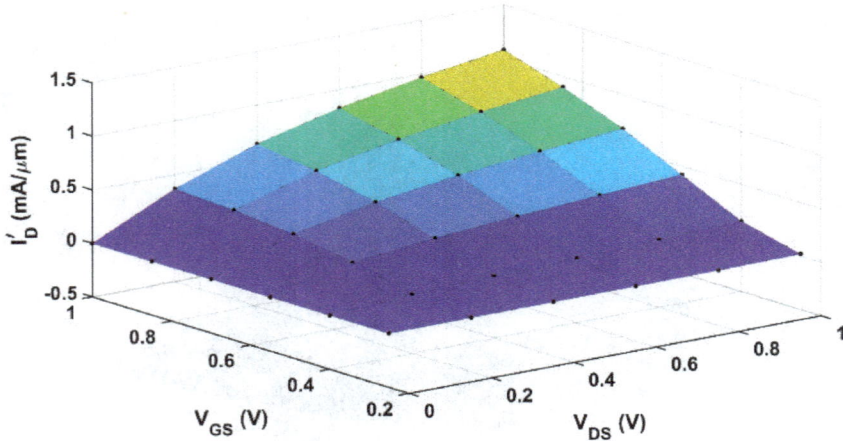

Fig. 7. Training Data set for model development (32 nm *n*-MOSFET).

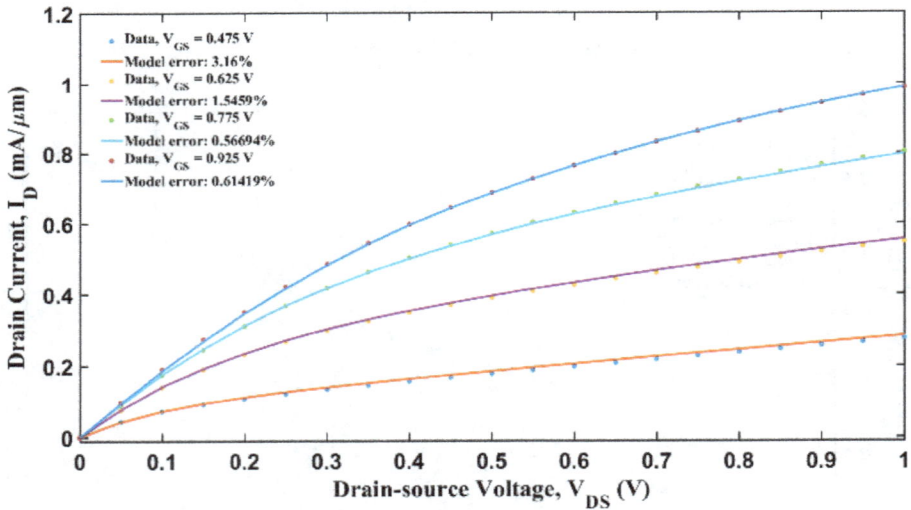

Fig. 8. Comparison between model and data for 32 nm *n*-MOSFET.

4.1.2.4.   45 nm Channel Length

To demonstrate the range of validity of the model, simulations have also been performed for 45 nm channel length NMOS device. The training data set is shown in Fig. 9 and the results for a different test set of gate voltage is shown in Fig. 10. As seen from these figures (Fig. 1 to Fig. 10), the model is capable of quite accurately predicting the MOSFET DC characteristics from long-channel devices observing square law to deep sub-micron devices with severe short-channel effects.

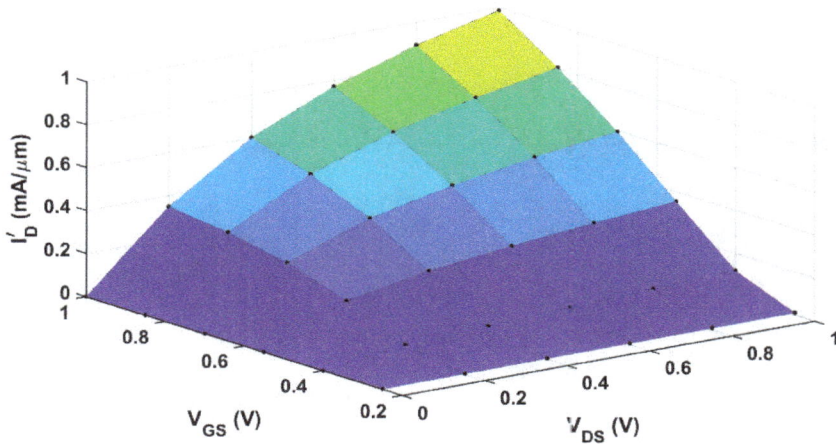

Fig. 9. Training data set for model development (45 nm *n*-MOSFET).

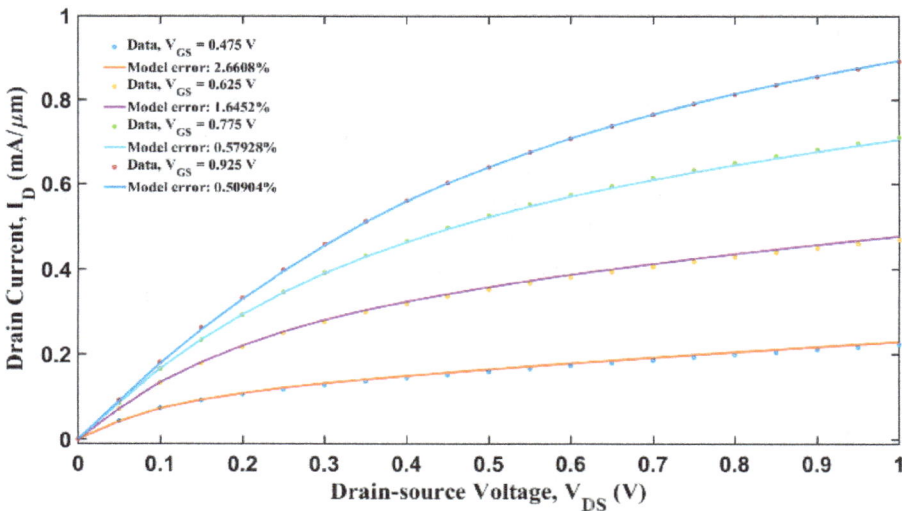

Fig. 10. Comparison between model and data for 45 nm *n*-MOSFET.

4.2. *MESFET*

The Metal-Semiconductor-Field-Effect-Transistor (MESFET) consists of a conducting channel between a source and drain contact region. A Schottky metal gate controls the carrier flow by varying the depletion layer width underneath the metal contact which modulates the thickness of the conducting channel to modulate the current. The model for a long channel normally-on SOI-MESFET can be developed following the same procedure as described for long channel MOSFETs. The normalized drain current for the *n*-channel MESFET can be written as:

$$I'_{DS} = \frac{I_{DS}}{W/L}$$

$$= \begin{cases} 0 \; ; V_{GS} \leq V_{TH} \\ q\mu_n N_d d \left[V_{DS} - \frac{2}{3} \left(\frac{(\varphi_i - V_{GS} + V_{DS})^{\frac{3}{2}}}{\sqrt{V_p}} - \frac{(\varphi_i - V_{GS})^{\frac{3}{2}}}{\sqrt{V_p}} \right) \right] (1 + \lambda V_{DS}); V_{GS} - V_{TH} \geq V_{DS} \\ q\mu_n N_d d \left[V_{GS} - V_{TH} - \frac{2}{3} \left(V_p - \frac{(\varphi_i - V_{GS})^{\frac{3}{2}}}{\sqrt{V_p}} \right) \right] (1 + \lambda V_{DS}) \; ; V_{GS} - V_{TH} \leq V_{DS} \end{cases} \quad (9)$$

where, V_{DS} is the drain-to-source voltage, V_{GS} is the gate-to-source voltage, V_{TH} is device threshold voltage, W is device channel width, L is device channel length, μ_n is the mobility of electron, N_d is channel doping density, q is the charge of an electron, d is the thickness of the channel, ε is the electric permittivity of the channel material, φ_i is the built-in Schottky potential between gate metal and channel material and λ is the channel length modulation parameter. The values of the parameters used in the simulation are, $\mu_n = 1000$ cm^2/(V.s), $V_{TH} = -1.0579$ V, $\varphi_i = 0.8564$ V, $V_p = 1.9143$ V, $d = 49.79$ nm, $\lambda = 0.05$ V^{-1}.

Figure 11 shows the surface plot of I'_D as a function of V_{DS} and V_{GS}. The dots show the data points evaluated from equation (9). Parameters of the model i.e. regression coefficients are calculated from these data points following the procedure described in Sec. 3. Then the normalized drain current, I'_D for unknown drain and gate biasing condition are calculated using the multivariate regression polynomial extended for two variables. Now to check the accuracy of the developed model another set of data is generated from equation (9). These new values from equation (9) are compared to the values predicted by the model. Figure 12 shows the comparison between the results obtained from the developed model to those

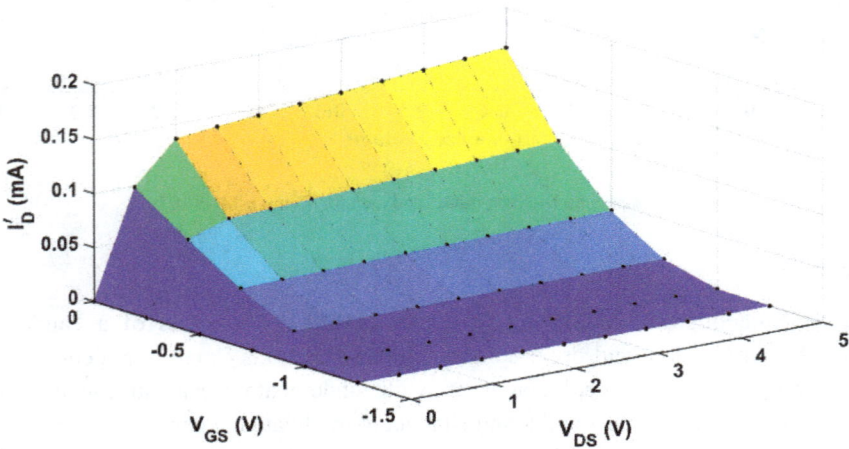

Fig. 11. Training data set for model development (n-MESFET).

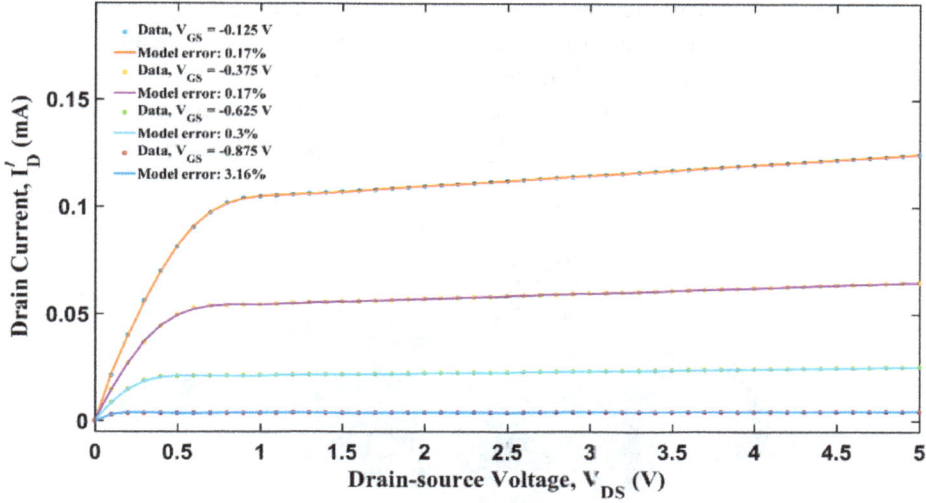

Fig. 12. Comparison between model and data for *n*-MESFET.

generated from equation (9). It shows that the values obtained from regression model match quite well with test data.

4.3. *HBT*

The heterojunction bipolar transistor (HBT) is a type of bipolar junction transistor (BJT) which uses different semiconductor materials for the emitter and the base regions, creating a heterojunction. The HBT has several improvements over the BJT such as it can handle signals of very high frequencies, up to several hundred GHz. It is commonly used in modern ultrafast circuits, mostly radio-frequency (RF) systems, and in applications requiring a high power efficiency, such as RF power amplifiers in cellular phones. Detailed theory of heterojunction bipolar transistor was developed by Herbert Kroemer in 1957 [3]. Here, the results for *n-p-n* and *p-n-p* HBT are shown and the results for conventional homojunction *n-p-n* BJT are also included. All simulations are performed in Sentaurus Device using the hydrodynamic transport model, where the carrier temperature equation for the dominant carriers – electrons for the *n-p-n* and holes for the *p-n-p* transistors – is solved together with the electrostatic Poisson equation and the carrier continuity equations. Doping-dependent mobility model is used, high-field saturation effects are accounted for and Shockley–Read–Hall and Auger generation–recombination models are activated.

4.3.1. *n-p-n Homojunction and Heterojunction BJT*

First, we show the results for an analytic $Si_{0.84}Ge_{0.16}$ *n-p-n* HBT device structure, which is created with the Sentaurus Structure Editor. The peak concentration of the base, emitter, and collector implant are defined as 5×10^{18}, 5×10^{19} and 1×10^{18} cm^{-3}, respectively. The training data set has been generated for six different base currents equally spaced between 1 μA/μm to 5 μA/μm and it consists of normalized collector current for collector-emitter voltage ranging from 0 to 1.5 V. Then the developed model is tested against a test data set for five different base currents equally spaced from 1.4 μA/μm to 4.6 μA/μm. The training data and test results for isolines are shown in Fig. 13 and Fig. 14, respectively.

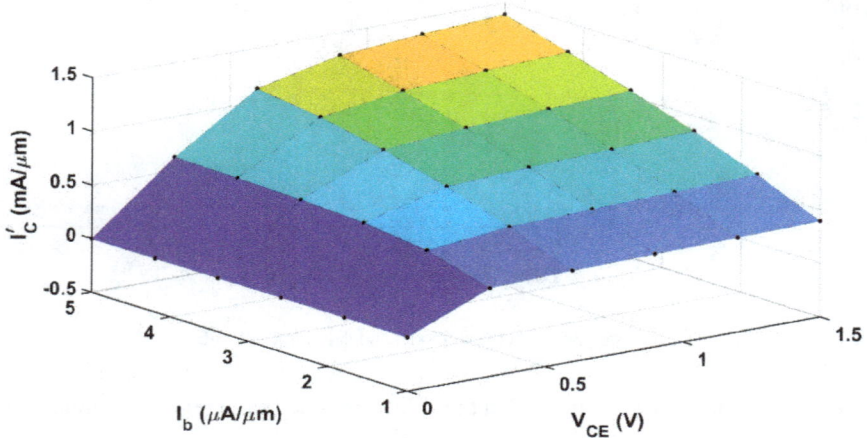

Fig. 13. Training data set for model development ($Si_{0.84}Ge_{0.16}$ *n-p-n* HBT).

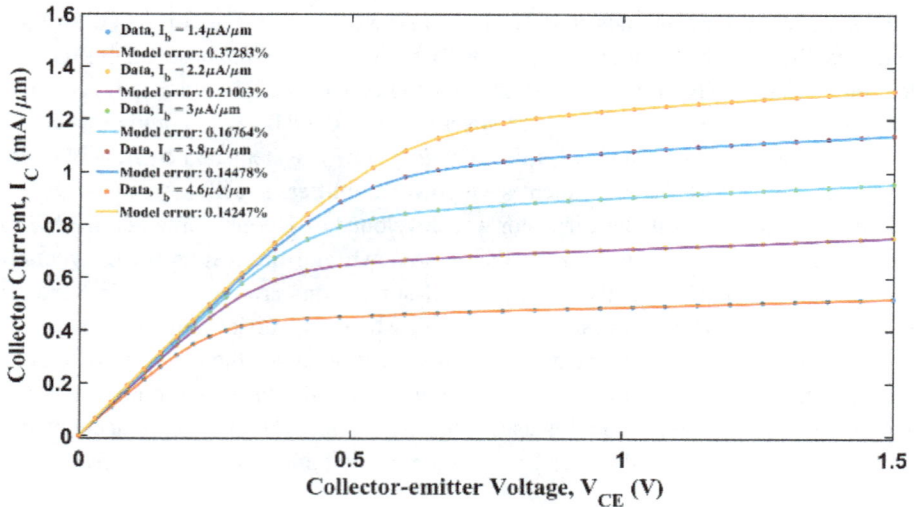

Fig. 14. Comparison between model and data for $Si_{0.84}Ge_{0.16}$ *n-p-n* HBT.

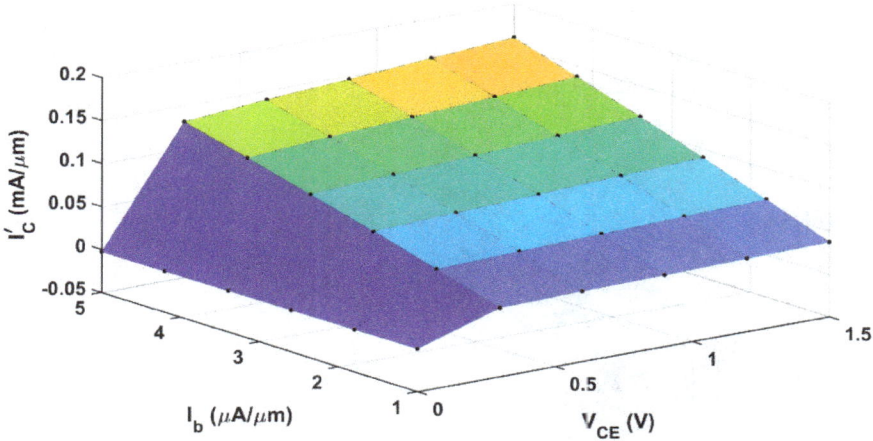

Fig. 15. Training data set for model development (Si *n-p-n* BJT).

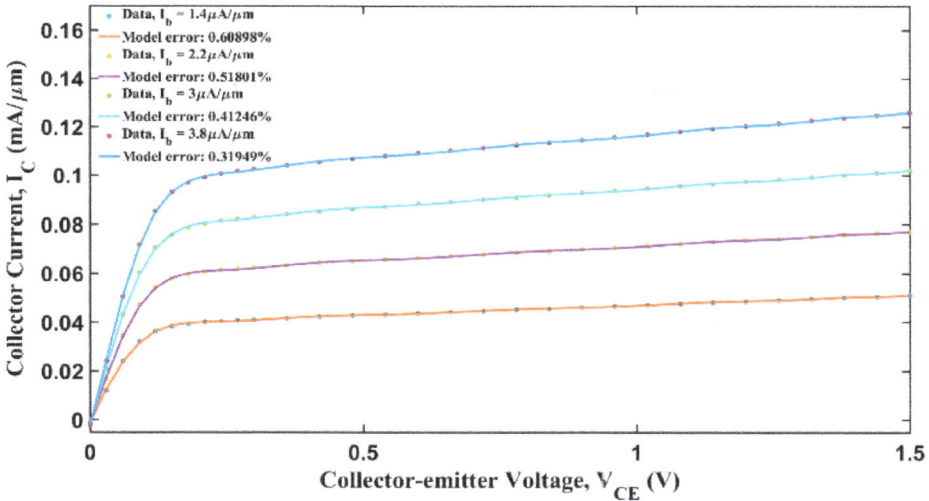

Fig. 16. Comparison between model and data for Si *n-p-n* BJT.

A conventional homojunction Si *n-p-n* BJT has also been modelled using regression polynomial. The training data set for this device is shown in Fig. 15 and the comparison between model prediction and a new set of test data is shown in Fig. 16. As shown in these figures, the model works almost equally well for both Si BJT and SiGe HBT transistors.

4.3.2. *p-n-p HBT*

For completeness, the results for *p-n-p* $Si_{0.84}Ge_{0.16}$ HBT have also been generated. The modeling follows the same procedure outlined in the previous section. The training data set is shown in Fig. 17 and the isolines with corresponding mean error is shown in Fig. 18. The model accuracy is quite good for both types of devices. Here, we show the magnitudes of normalized base and collector current.

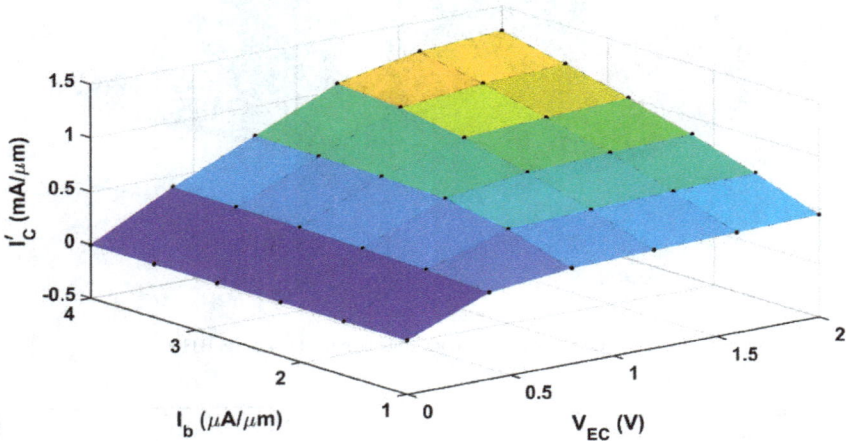

Fig. 17. Training data set for model development (Si$_{0.84}$Ge$_{0.16}$ *pnp*-HBT).

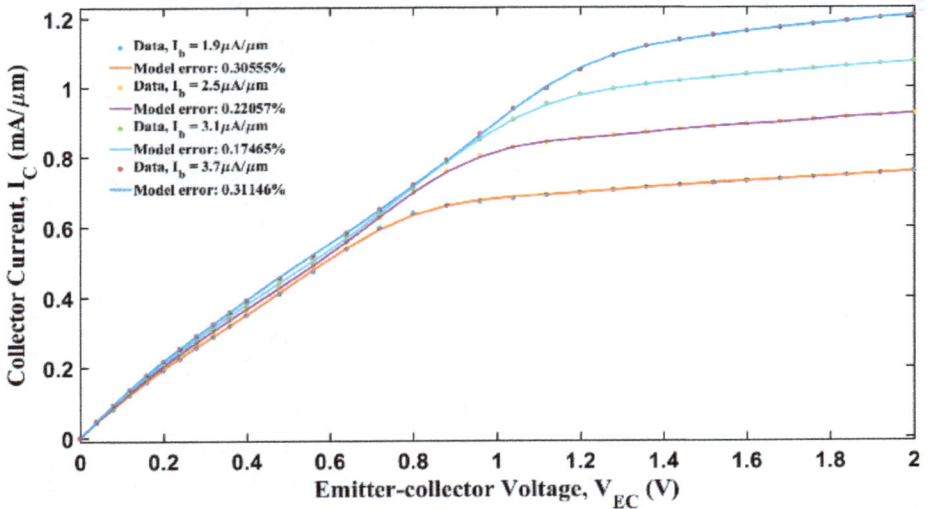

Fig. 18. Comparison between model and data for Si$_{0.84}$Ge$_{0.16}$ *pnp*-HBT.

4.4. *HEMT*

The multivariate regression polynomial model is used to model the DC I-V characteristics of InGaAs HEMTs [24, 25]. TCAD Sentaurus is used to generate both the training and the

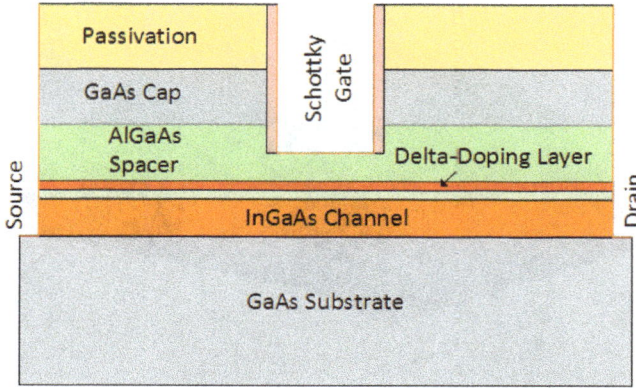

Fig. 19. Schematic of HEMT structure used for TCAD simulation.

test data sets. A schematic of the HEMT device structure used for TCAD simulation is shown in Fig. 19.

The parameters used in HEMT layer structure are as follows: Substrate = 0.8 μm GaAs, channel = 10 nm InGaAs, spacer = 34.5 nm AlGaAs, cap layer = 30 nm GaAs, passivation layer = 50 nm nitride, location of delta-doping layer = 31 nm, thickness of doping layer = 2 nm. The Schottky gate contact is etched into the top spacer layer. This gate recess is 15 nm deep and gate length is 0.25 μm. At each side of the gate, a 40 nm wide oxide layer isolates the gate from the cap layer. The sheet doping concentration is 5.4×10^{12} cm^{-2}. It is assumed that the dopants are in a tight Gaussian distribution, with a diffusion length given by the thickness of the delta-doping layer. The mole fraction in the $In_{(1-x)}Ga_xAs$ channel is set to x = 0.75. The mole fraction in the $Al_xGa_{(1-x)}As$ spacer layer is set to x = 0.3. A donor trap concentration of 10^{16} cm^{-3} is set in the substrate layer. Hydrodynamic transport model is used for the electrons with both high field mobility saturation model and impact ionization model activated. Furthermore, Shockley–Read–Hall (SRH), Auger, and radiative recombination models are activated. At the Ohmic source and the drain contacts, contact resistances of 150 Ωμm are set and the gate is defined as a Schottky contact with a Schottky barrier height of 0.9 eV and electron and hole recombination velocities of 10^7 cm/s.

The training data set has been generated for six different gate voltages ranging from -0.8 V to 0.8 V in 0.32 V increment. It consists for normalized drain current for drain voltage ranging from 0 to 1.5 V. Figure 20 shows the surface plot of the training data. Then the developed model is tested against a test data set for five different gate voltages ranging from -0.64 V to 0.64 V in 0.32 V increment. The test results for isolines are shown in Fig. 21. The worst case error is 0.63% for gate voltage of -0.64 V. The small relative errors for all five cases show excellent match between TCAD data and multivariate regression model.

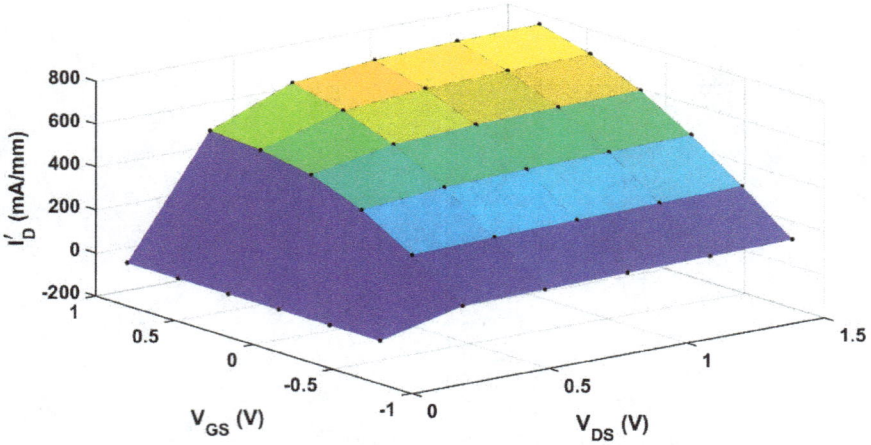

Fig. 20. Training data set for model development (InGaAs HEMT).

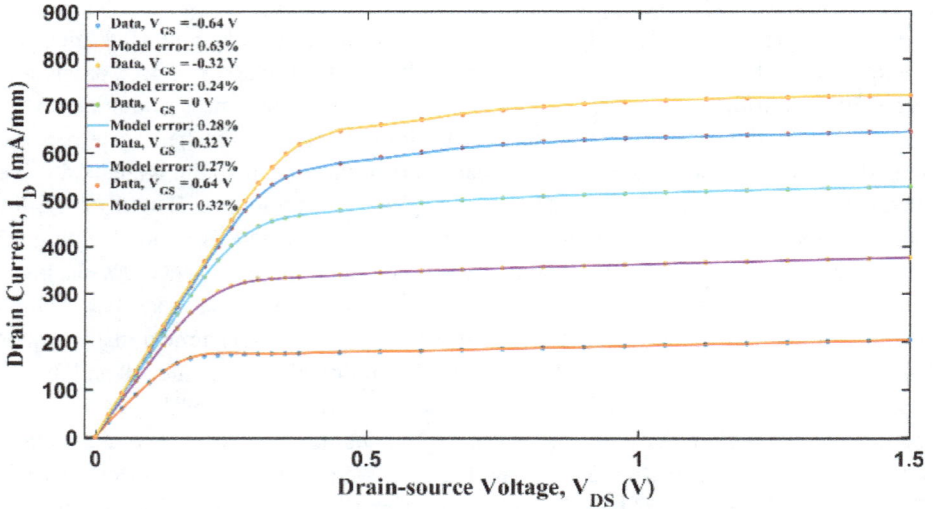

Fig. 21. Comparison between model and data for InGaAs HEMT.

4.5. *G⁴FET*

This model can be used to include multi-gates into the model development with relative ease while still being capable of predicting DC response quite accurately. As an example, a four gate silicon-on-insulator transistor known as G⁴FET has been chosen. G⁴FET is a relatively recent SOI (silicon-on-insulator) device ([5], [6]) which works as a four-gate transistor with two lateral junction-gates (JFET action) and two vertical oxide gates (MOS action). The top oxide gate is the conventional MOS gate whereas the buried oxide along with the substrate biasing is used for bottom oxide gate and they are used to control the accumulation/depletion/inversion of free carriers near the top and the bottom oxide interfaces, respectively. An *n*-channel G⁴FET can be constructed from conventional *p*-channel SOI MOSFET with two body contacts on the opposite sides of the channel. The source and the drain of a *p*-channel MOSFET are *p*+ doped, and act as lateral junction-gates providing JFET action of controlling the width of the conduction channel, whereas, the body contacts are used as the source and the drain for the *n*-channel G⁴FET. Thus, an inversion-mode, *p*-channel MOSFET is turned into an accumulation/depletion-mode *n*-channel G⁴FET.

Few examples of experimentally demonstrated innovative circuit applications using this device are analog multipliers [26], temperature compensated voltage references [27], digital applications such as multi-threshold inverter, universal and programmable gate, three transistor full adder [28] etc. A simple two transistor tunable negative resistance circuit built with G⁴FETs has been used to realize LC oscillators and Schmitt trigger circuits [29] and its tent map shape characteristics can be leveraged to build chaotic logic gates suitable for hardware security applications [30]. G⁴FET inspired multiple state electrostatically formed nanowires have also been used for threshold logic functions [31, 32] and high sensitivity gas sensing and femtomolar bio-marker detection [33, 34].

To capture the current-voltage (*I-V*) characteristic of a G⁴FET, regression polynomial was used to derive the numerical model from the available data set. Current-voltage data for different bias values have been gathered from both experiment and TCAD Sentaurus for both *p*-channel and *n*-channel G⁴FET transistors. Then a polynomial regression model is chosen consisting of suitable model terms and the coefficients of this model are estimated for least square error. The total number of terms in the model is chosen by user and it also

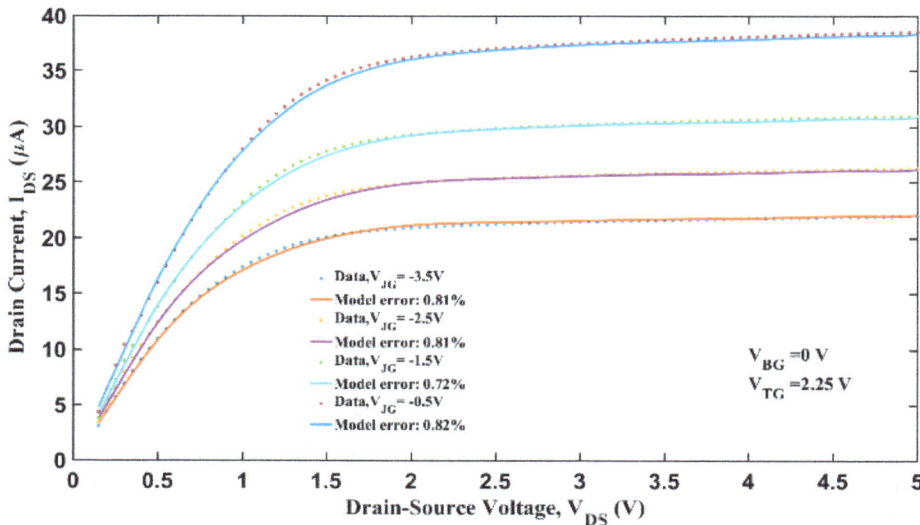

Fig. 22. Comparison between model and TCAD data for long-channel *n*-MOSFET.

allows non-integer exponents. The current-voltage characteristics predicted by the model are then tested against another set of test data. Here, we briefly show the results for two cases. The first case consists of an *n*-channel G^{4}FET simulated with TCAD Sentaurus. The training data set has been generated for different drain-source (V_{DS}), top gate (V_{TG}), bottom-gate (V_{BG}) and junction gate (V_{JG}) voltages. For simplicity, both junction gates are tied together and then based on these data, drain current (I_{DS}) regression model is developed as a function of four voltages (V_{DS}, V_{TG}, V_{BG}, V_{JG}). The test results for different top gate voltages is shown in Fig. 23. To demonstrate the predictive power of the model, we have also used experimental measurement from a PDSOI *n*-channel G^{4}FET with width and length of 0.4 µm and 0.9 µm, respectively to develop a regression model. Then the model is tested against a new set of test data is shown in Fig. 24. The corresponding mean error for different operating conditions show quite accurate match between the experimental data and the model. A more detailed exposition on this work has been reported in [20].

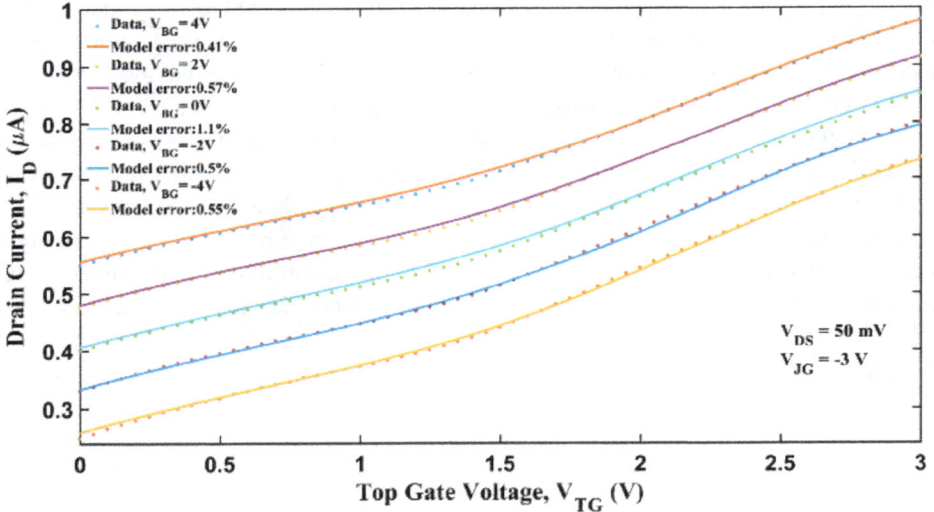

Fig. 23. Comparison between model and experimental data for long-channel *n*-MOSFET.

5. Conclusion

In this work, a multivariable regression polynomial model is used to develop DC model of five different modern transistors and subsequently validated using analytical results, TCAD data and experimental results illustrating its accuracy and versatility. The model with its first and second order continuity and a single polynomial expression valid in all operating regimes is suitable for circuit simulator implementation. Although the independent variables considered in this work include terminal voltages/currents, the same procedure can be used to extend the model by incorporating more variables such as geometric dimensions, terminal capacitances etc. for scalable transient and ac simulation.

As demonstrated with five transistors with very different physical conduction mechanisms, the developed model is quite general and can be used in modeling any emerging single/multi-gate device for which accurate physics based simplified compact model is either too complex to be efficiently evaluated or has not yet been developed.

References

1. P. Antognetti and G. Massobrio, "Semiconductor Device Modeling With Spice", New York: McGraw-Hill, 1988.
2. J. Ervin, A. Balijepalli, P. Joshi, J. Yang, V. Kushner, and T. J. Thornton, "CMOS-Compatible SOI MESFETs With High Breakdown Voltage", in *IEEE Transactions on Electron Devices*, vol. 53, pp. 3129-3135, 2006.
3. H. Kroemer, "Theory of a Wide-Gap Emitter for Transistors", in *Proceedings of the IRE*, vol. 45, no. 11, pp. 1535-1537, Nov. 1957.
4. T. Mimura, "The early history of the high electron mobility transistor (HEMT)", in *IEEE Transactions on Microwave Theory and Techniques*, vol. 50, no. 3, pp. 780-782, March 2002.
5. S. Cristoloveanu, B. Blalock, F. Allibert, B. Dufrene and M. Mojarradi, "The Four-Gate Transistor", in *Proceedings of the 2002 European Solid-State Device Research Conference*, pp. 323-326, Firenze, Italy, September 2002.
6. B. J. Blalock, S. Cristoloveanu, B. M. Dufrene, F. Allibert, and M. M. Mojarradi, "The Multiple-Gate MOS-JFET Transistor", *International Journal of High Speed Electronics and Systems*, vol. 12, no. 2, pp. 511-520, 2002.
7. A. Rofougaran and A. A. Abidi, "A table lookup FET model for accurate analog circuit simulation", in *IEEE Transactions on Computer-Aided Design of Integrated Circuits and Systems*, vol. 12, no. 2, pp. 324-335, Feb 1993.
8. A. R. Rofougaran, B. Furman and A. A. Abidi, "Accurate analog modeling of short channel FETs based on table lookup", in *Proceedings of the IEEE 1988 Custom Integrated Circuits Conference*, Rochester, NY, 1988, pp. 13.1/1-13.1/4.
9. Schrom, Gerhard, Andreas Stach, and Siegfried Selberherr, "An interpolation based MOSFET model for low-voltage applications", *Microelectronics journal* 29, no. 8 (1998): 529-534.
10. P. B. L. Meijer, "Fast and smooth highly nonlinear multidimensional table models for device modeling", in *IEEE Transactions on Circuits and Systems*, vol. 37, no. 3, pp. 335-346, Mar 1990.
11. M. G. Graham and J. J. Paulos, "Interpolation of MOSFET table data in width, length, and temperature", in *IEEE Transactions on Computer-Aided Design of Integrated Circuits and Systems*, vol. 12, no. 12, pp. 1880-1884, Dec 1993.
12. V. Bourenkov, K. G. McCarthy, and A. Mathewson, "MOS table models for circuit simulation", in *IEEE Transactions on Computer-Aided Design of Integrated Circuits and Systems*, vol. 24, no. 3, pp. 352-362, Mar 2005.
13. M. Yanilmaz and V. Eveleigh, "Numerical device modeling for electronic circuit simulation", in *IEEE Transactions on Computer-Aided Design of Integrated Circuits and Systems*, vol. 10, no. 3, pp. 366-375, Mar 1991.
14. M. S. Hasan, T. Rahman, S. K. Islam, and B. J. Blalock, "Numerical modeling and implementation in circuit simulator of SOI four-gate transistor (G4FET) using multidimensional Lagrange and Bernstein polynomial", in *Microelectronics Journal*, vol. 65, pp. 84-93, July 2017.

15. T. Shima, H. Tamada, Ryo Luong and Mo Dang, "Table Look-Up MOSFET Modeling System Using a 2-D Device Simulator and Monotonic Piecewise Cubic Interpolation", in *IEEE Transactions on Computer-Aided Design of Integrated Circuits and Systems*, vol. 2, no. 2, pp. 121-126, April 1983.
16. B. J. Burns, "Empirical MOSFET models for circuit simulation", Memo M84/43, Electron. Res. Lab., Univ. California, Berkeley, 1984.
17. J. A. Barby, J. Vlach, and K. Singhal, "Polynomial splines for MOSFET model approximation", in *IEEE Transactions on Computer-Aided Design of Integrated Circuits and Systems*, vol. 7, no. 5, pp. 557-566, May 1988.
18. M. S. Hasan, S. K. Islam, and B. J. Blalock, "Modeling of SOI four-gate transistor (G 4 FET) using multidimensional spline interpolation method", in *Microelectronics Journal,* vol. 76, pp. 33-42, Jun 2018.
19. P. B. L. Meijer, "Table models for device modelling", in *1988, IEEE International Symposium on Circuits and Systems*, Espoo, 1988, pp. 2593-2596 vol. 3.
20. M. S. Hasan and S. K. Islam, "DC Modeling of SOI Four-Gate Transistor (G4FET) for Implementation in Circuit Simulator Using Multivariate Regression Polynomial", in *IET Circuits, Devices & Systems*, Apr 2018.
21. D'Errico, John, 'Polyfitn', MATLAB Central File Exchange. Retrieved August 6, 2015.
22. H. Shichman and D. A. Hodges. "Modeling and simulation of insulated-gate field-effect transistor switching circuits", in *IEEE J. Solid State Circuits*, SC-3, 1968.
23. W. Shockley, "Circuit Element Utilizing Semiconductive Material", United States Patent 2,569,347, 1951.
24. K. H. G. Duh, P. C. Chao, S. M. J. Liu, P. Ho, M. Y. Kao and J. M. Ballingall, "A super low-noise 0.1 μm T-gate InAlAs-InGaAs-InP HEMT", in *IEEE Microwave and Guided Wave Letters*, vol. 1, no. 5, pp. 114-116, May 1991.
25. T. Akazaki, K. Arai, T. Enoki and Y. Ishii, "Improved InAlAs/InGaAs HEMT characteristics by inserting an InAs layer into the InGaAs channel", in *IEEE Electron Device Letters*, vol. 13, no. 6, pp. 325-327, June 1992.
26. K. Akarvardar, S. Chen, B.J. Blalock, S. Cristoloveanu, P. Gentil, M. Mojarradi, "A Novel Four-Quadrant Analog Multiplier Using SOI Four-Gate Transistors (G4-FETs)", in *Proceedings of ESSCIRC*, Grenoble, France, 2005.
27. S. C. Terry, S. Chen, B. J. Blalock, J. R. Jackson, B. M. Dufrene, M. M. Mojarradi, S. K. Islam and M. N. Ericson, "Temperature-Compensated Reference Circuits for SOI", *2004 IEEE International SOI Conference*, October 2004.
28. K. Akarvardar, B. Blalock, S. Chen, S. Cristoloveanu, P. Gentil and M. M. Mojarradi, "Digital Circuits Using SOI Four-Gate Transistor," *2006 8th International Conference on Solid-State and Integrated Circuit Technology Proceedings*, Shanghai, 2006, pp. 1867-1869.
29. K. Akarvardar, S. Chen, J. Vandersand, B. Blalock, R. Schrimpf, B. Prothro, C. Britton, S. Cristoloveanu, P. Gentill, and M. M. Mojarradi, "Four-Gate Transistor Voltage-Controlled Negative Differential Resistance Device and Related Circuit Applications", *2006 IEEE International SOI Conference Proceedings*.
30. M. B. Majumder, M. S. Hasan, M. Uddin and G. S. Rose, "Chaos computing for mitigating side channel attack", in *2018 IEEE International Symposium on Hardware Oriented Security and Trust (HOST)*, Washington, DC, 2018, pp. 143-146.

31. G. Segev, I. Amit, A. Godkin, A. Henning, and Y. Rosenwaks, "Multiple State Electrostatically Formed Nanowire Transistors", in *IEEE Electron Device Letters*, vol. 36, no. 7, pp. 651-653, July 2015.

32. J. S. Friedman, A. Godkin, A. Henning, Y. Vaknin, Y. Rosenwaks and A. V. Sahakian, "Threshold Logic with Electrostatically Formed Nanowires." in *IEEE Transactions on Electron Devices*, vol. 63, no. 3, pp. 1388-1391, March 2016.

33. G. Shalev, G. Landman, I. Amit, Y. Rosenwaks, and I. Levy, 'Specific and label-free femtomolar biomarker detection with an electrostatically formed nanowire biosensor', *NPG Asia Mater.*, 5(3), p. e41, 2013.

34. A. Henning, N. Swaminathan, A. Godkin, G. Shalev, I. Amit, and Y. Rosenwaks, "Tunable diameter electrostatically formed nanowire for high sensitivity gas sensing", *Nano Research*, vol. 8, no. 7 (2015): 2206-2215.

A SPICE Model for GaN-Gate Injection Transistor (GIT) at Room Temperature

Frances Garcia[1,*], Samira Shamsir[2], Syed K. Islam[1,2] and Leon M. Tolbert[1]

[1]Min. H. Kao Department of Electrical Engineering and Computer Science,
The University of Tennessee, Knoxville, TN 37996, USA
*fgarcia@vols.utk.edu

[2]Department of Electrical Engineering and Computer Science,
University of Missouri, Columbia, MO 65211, USA

In this paper, an equivalent circuit model is developed for a commercial Gallium Nitride (GaN) gate injection transistor (GIT) device at room temperature. The *I-V* and *C-V* characteristics are extracted from the commercial device datasheet and fitted in MATLAB. The fitted equations are realized as a combination of behavioral circuit components to be carried out in SPICE simulation. The equivalent circuit model is tested in a simple configuration for *I-V* curve simulation through varying of parameters and then in a simple boost configuration typical for power electronics applications to check for convergence and expected results.

Keywords: GaN-GIT; SPICE model; boost converter.

1. Introduction

GaN poses as a good candidate for power devices due to its ability to sustain high power density and high frequency, and power devices based on this material system are becoming an attractive option for power electronics. For power applications, the transistors are preferred to be enhancement mode for safety reasons. There are only a few available GaN enhancement mode devices commercially, but most tend to be recessed high electron mobility transistors (HEMTs)[1]. The device studied in this paper is a Panasonic gate injection transistor (GIT)[2] whose data sheet is publicly available. While the recessed HEMT operates by reducing the AlGaN barrier thickness and thereby reducing the 2DEG in the channel, the gate injection transistor has a *p*-doped GaN layer below the gate electrode which raises the conduction band above the Fermi level turning the device off at zero gate bias[3]. The typical structure of a GIT is shown in Fig. 1.

As the gate voltage is increased beyond the threshold voltage, the Fermi level is lowered below the conduction band, turning the device on. Further increase in the gate voltage causes holes to be injected into the channel inducing conductivity modulation[4]. Electrons are pulled as the drain-to-source voltage is increased while the holes stay near the gate region due to their lower mobility. Empirical models of GITs or other GaN power devices have been reported in the literature. However, these are difficult to implement into

SPICE syntax with smooth transition, reverse conduction, and convergence or it may be implemented by a BSIM model which may not capture the characteristics of the GIT well[5,6]. This paper focuses on a Panasonic GIT that is modeled using SPICE friendly syntax which includes a smooth transition from linear to saturation regions, reverse conduction, and fast convergence in SPICE simulators.

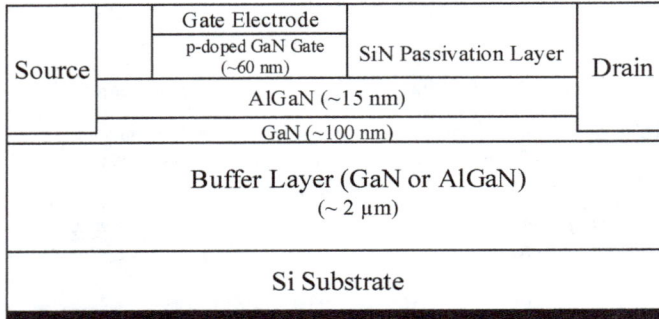

Fig. 1. Typical structure for the GaN gate injection transistor[4].

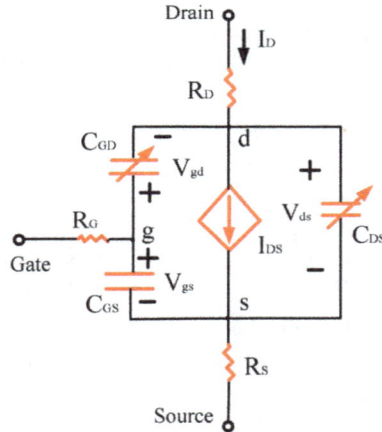

Fig. 2. The equivalent circuit model for the GIT including parasitic resistance and capacitance[7].

2. Development of the Model

The commercial device modeled in this paper is the Panasonic PGA26E07BA device. Based on the datasheet, the *I-V* and *C-V* characteristics were extracted. The extracted data was fitted using MATLAB™ software to easily implement into a simple SPICE simulator with an equivalent circuit model. Figure 2 shows the equivalent circuit model used with parasitic effects extracted from the *I-V* and *C-V* curves. The SPICE circuit used for the model as shown in Fig. 2 uses a voltage-dependent current source, I_{DS}, three voltage-dependent capacitances, and three constant parasitic resistances. The parasitic resistances used were directly obtained from the datasheet. The voltage-dependent current source is used to model static current-voltage characteristics for both forward and reverse conduction mode possibilities in GaN devices. The parasitic resistance and capacitances are necessary to provide insight into the behavior of switching performance of the device. The voltage-dependent current source is a bidirectional current source used to mimic the reverse conduction behavior of GaN. The current source is fitted in MATLAB and treated as a voltage-dependent switched current source.

The parasitic capacitance was included as a drain to source voltage-dependent nonlinear capacitance extracted from the datasheet. The sheet includes the capacitance measurements for C_{oss}, C_{rss}, and C_{iss} which are important parameters in the realm of switching power electronics. In order to make this information useful for the equivalent model, eqns. (1)-(3) were used to convert the parameters into useful C_{GD}, C_{GS}, and C_{DS}.

$$C_{iss} = C_{GD} + C_{GS} \tag{1}$$

$$C_{oss} = C_{DS} + C_{GD} \tag{2}$$

$$C_{rss} = C_{GD} \tag{3}$$

The parasitic resistances are assumed to be constant and are obtained from the data sheet.

3. Extraction and Fitting

I-V and *C-V* curves were extracted and imported into MATLAB for a logarithmic fit. While a polynomial fit or compact model could have been used, it is difficult to implement these models into a SPICE environment with the inclusion of a smooth transition from the linear to the saturation region as well as for reverse conduction. Figure 3 shows the extracted *I-V* curves with the surface fit in MATLAB dependent on V_{GS} and V_{DS}. The 2D fit in Fig. 4 gives better insight into the discrepancy in the linear region. The saturation region most closely matches the device experimental data which is the region where the device is typically used for switching applications such as DC-DC converters. The fitted equation is given in Fig. 3. While a better fit may be achieved with polynomial fits, they present problems that make it difficult to implement with SPICE syntax. Other disadvantages include the lack of reverse conduction, oscillations outside of the fitted data, and nonzero convergence at $V_{DS} = 0$ V for values outside of the ranges the interpolated fit is based on. The GIT model provides a general behavioral model for circuit simulation using the logarithmic fit. Logarithmic functions can be easily implemented into a multivariable voltage-dependent current source, but are difficult to adjust in the linear region with fitting parameters. Due to the nature of the logarithmic fit, the model efficiency remains an issue in circuit simulation, but provides insight into the device behavior for general circuit modeling.

It is necessary to have *C-V* curves in a model because many of the applications for the gate injection transistor include switching applications such as DC-DC converters and

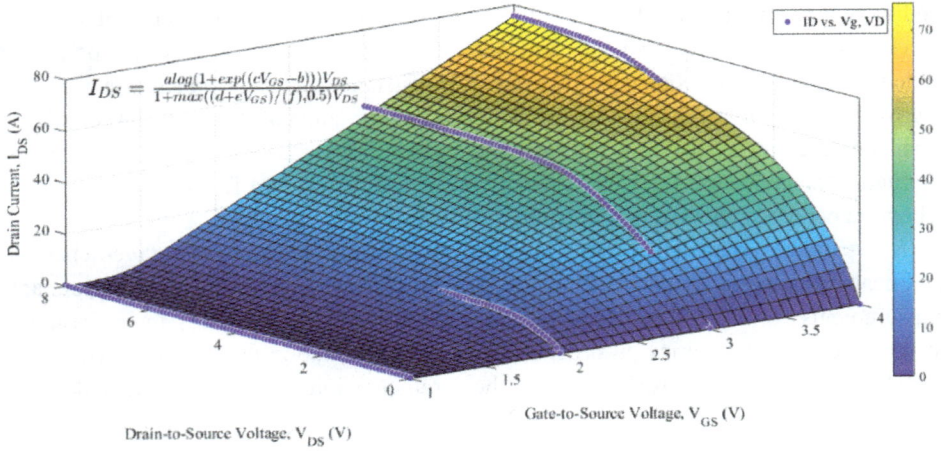

Fig. 3. 3D surface fitting in MATLAB for the GIT. The gate-to-source and drain-to-source voltages are varied with a fitting equation expression in the left-hand corner.

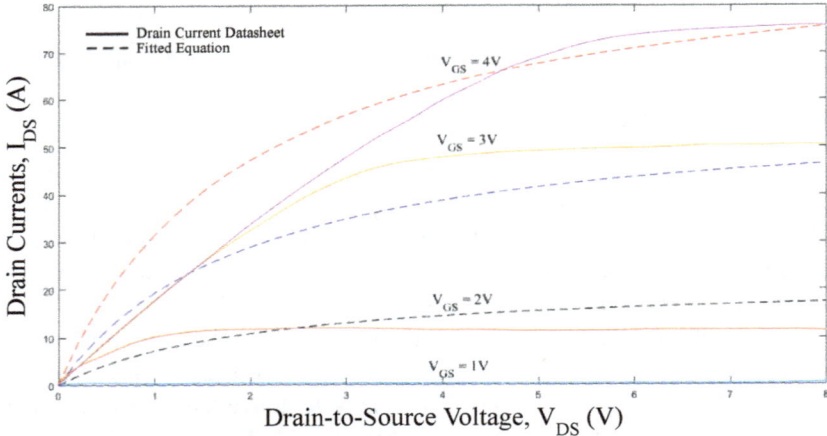

Fig. 4. 2D representation of the curve fitted equation.

inverters. The C-V curves were extracted and fitted with exponential expressions. The C-V curves were more closely matched and are shown in Figs. 5-7 for C_{oss}, C_{iss}, and C_{rss}. C_{oss} represents small signal output capacitance when gate and source terminals are shorted. C_{iss} is the small signal input capacitance when drain and source terminals are shorted, and C_{rss} is the small signal reverse transfer capacitance. Recalling eqns. (1)-(3), the fitted equations for C_{GS}, C_{GD}, and C_{DS} are acquired and are shown in eqns. (4)-(6).

$$C_{oss} = 187e^{-0.0116V_{DS}} + 82.72e^{-0.0004V_{DS}} \qquad (4)$$

$$C_{iss} = 145.8e^{-0.2098V_{DS}} + 405.2e^{-4.3x10^{-6}V_{DS}} \tag{5}$$

$$C_{rss} = 110.5e^{-0.241V_{DS}} + 32.54e^{-0.0187V_{DS}} \tag{6}$$

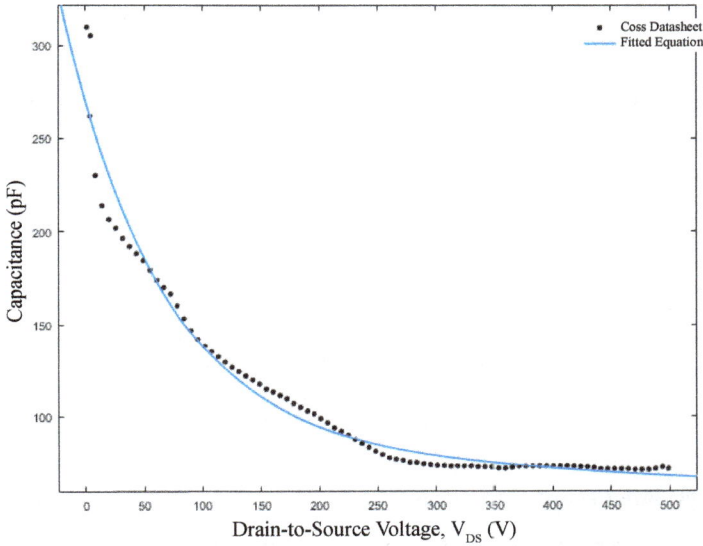

Fig. 5. C_{oss} fit from datasheet.

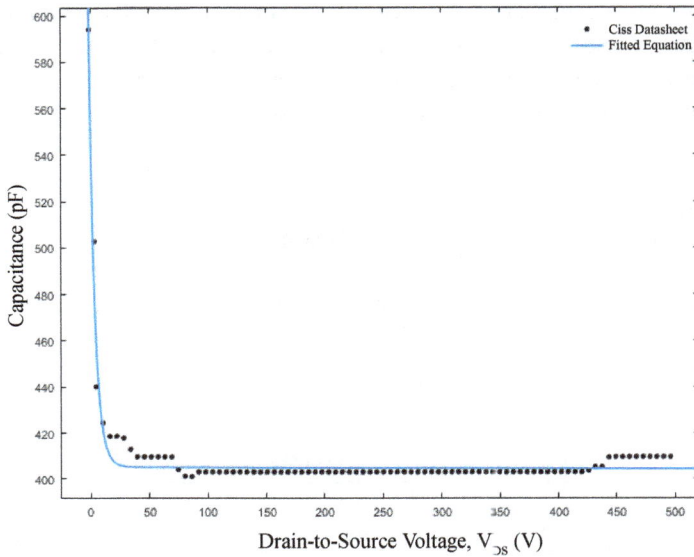

Fig. 6. C_{iss} fit from datasheet.

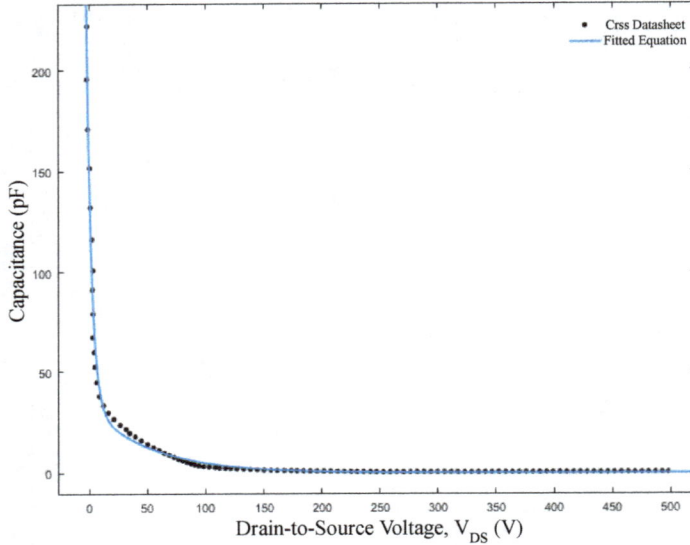

Fig. 7. C$_{rss}$ fit from datasheet.

4. SPICE Simulated Boost Converter

Once the equivalent circuit model was implemented as a sub-circuit in LTspice, it can be used as a part of larger more complex circuits for power applications. To test convergence and the functioning of the model, the sub-circuit was implemented into two test simulations. The first simulation in LTspice (a freeware computer software implementing a SPICE simulator of electronic circuit) was a simple varying of the gate-to-source and the drain-to-source voltages for checking convergence time and accuracy. Figure 8 shows the simulated circuit, and Fig. 9 shows the simulation results.

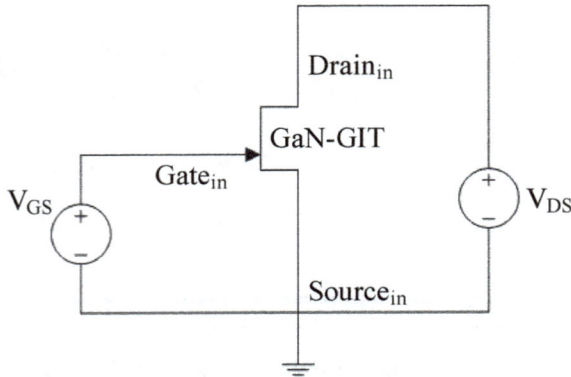

Fig. 8. LTspice schematic design for testing GIT equivalent model.

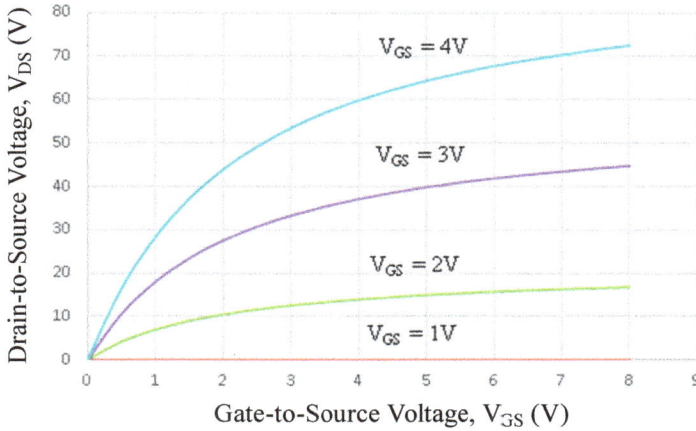

Fig. 9. LTspice simulation results of voltage parameter variation for Fig. 8.

The second circuit that was tested was a boost converter configuration. Usually complex empirical models can slow down the simulation time or cause the simulation to not converge at all. The fast switching implemented in the circuit allows for a more rigorous simulation test. Figure 10 shows the test circuit for a boost converter with an input of 25 V to 50 V output, switching frequency of 150 kHz, 3 A current ripple and 1 V ripple voltage. Figure 10 also includes a simple simulation command for a transient simulation for the previously mentioned parameters. The resulting waveforms for input current, inductor current, capacitor current, and output current are shown in Figs. 11-13.

The simulation results of input current, inductor current, capacitor current, and output current are obtained for a transient simulation time between 4-4.5 ms for better waveform resolution. The simulation results prove that the model converges with expected results and can be used as a guide for circuit behavior simulation.

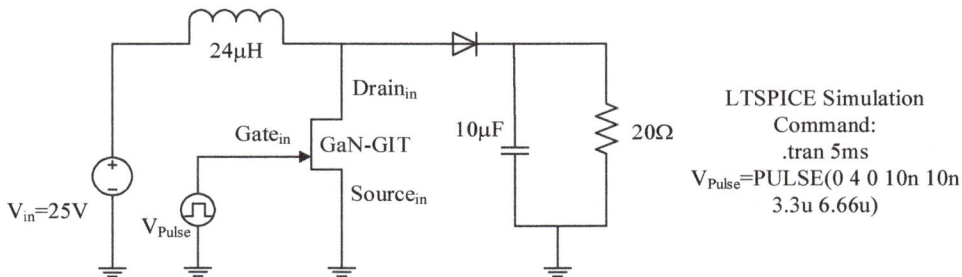

Fig. 10. Boost converter topology implemented in LTspice for transient boost simulation. The simulation command shows a switching frequency of 150 kHz, gate voltage maximum of 4 V and, and transient of 5 ms.

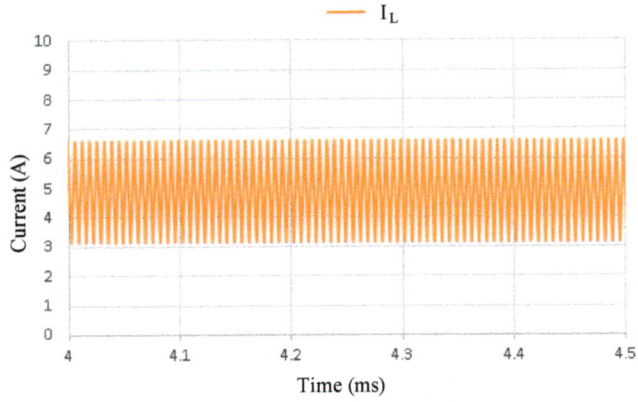

Fig. 11. LTspice simulation results of inductor current (I_L) for the boost converter topology from Fig. 10.

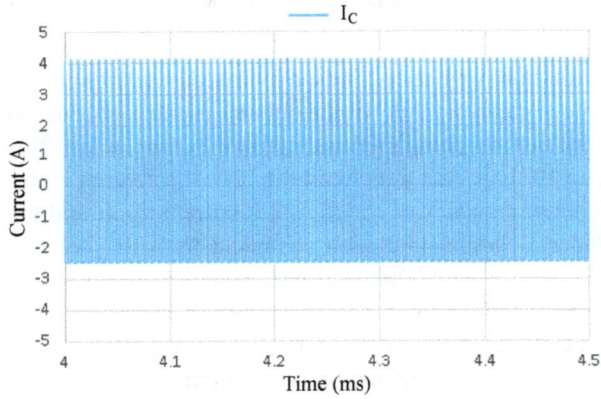

Fig. 12. LTspice simulation results of capacitor current for boost converter topology shown in Fig. 10.

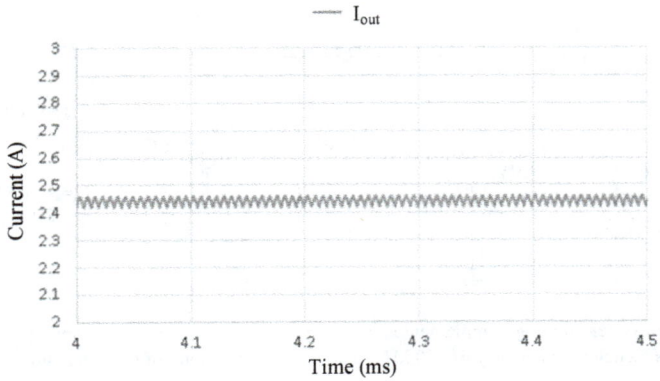

Fig. 13. LTspice simulation results of output current for boost converter topology shown in Fig. 10.

5. Conclusion

A simple equivalent circuit for the GaN gate injection transistor was developed for SPICE circuit simulator. The device behavior was modeled by extracting the drain current from static *I-V* and *C-V* characteristics. The device model captures reverse conduction, a smooth transition from linear to saturation region, and parasitic device parameters present at room temperature. Future work for the model includes introducing temperature dependence, improving efficiency modeling, and considering implementing the model into a standardized GaN HEMT compact model if the Compact Model Coalition (CMC) agrees upon a model for GaN HEMT[8,9]. The model achieves expected results in DC-DC converter simulation tests with quick convergence due to the compactness of the equivalent circuit.

References

1. E. A. Jones, F. Wang, and D. Costinett, "Review of commercial GaN power devices and GaN based converter design challenges," *IEEE Journal of Emerging and Selected Topics in Power Electronics* **4**, 707-719 (2016).
2. Panasonic, GaN-Tr N-channel enhancement mode FET, PGA26E07BA datasheet, Sept. 2016 (Revised Jan. 2017).
3. Z. Wang, B. Zhang, W. Chen, and Z. Li, "A closed-form charge control model for the threshold voltage of depletion and enhancement-mode AlGaN/GaN devices," *IEEE Transactions on Electron Devices* **60**(5), 1607-1612 (2014).
4. Y. Uemoto, M. Hikita, H. Ueno, H. Matsuo, H. Ishida, M. Yanagihara, T. Ueda, T. Tanaka, and D. Ueda, "Gate injection transistor (GIT) – A normally-off AlGaN/GaN power transistor using conductivity modulation," *IEEE Transactions on Electron Devices*, **54**(12), 3393-3399 (2007).
5. R. Kotecha, Y. Zhang, A. Rashid, N. Zhu, T. Vrostsos, and H. Mantooth, "A physics-based compact gallium nitride power semiconductor device model for advanced power electronics design," *IEEE Applied Power Electronics Conference and Exposition* (2017).
6. R. Kotecha, Y. Zhang, A. Wallace, N. Zhu, A. Rashid, T. Vrotsos, and H. Mantooth, "An accurate compact model for gallium nitride gate injection transistor for next generation power electronics design," *IEEE 18th Workshop on Control and Modeling for Power Electronics* (2017).
7. K. Peng, S. Eskandari, and E. Santi, "Characterization and Modeling of Gallium Nitride Power HEMT," *IEEE Transactions on Industry Applications*, **52**(6), 4965-4975 (2016).
8. S. D. Mertens, "Status of the GaN HEMT standardization effort at the compact model coalition," *IEEE Compound Semiconductor Integrated Circuit Symposium* (2014).
9. S. Khandelwal, C. Yadav, S. Agnihotri, Y. S. Chauhan, A. Curutchet, T. Zimmer, J. Jaeger, N. Defrance, and T. A. Fjeldly, "Robust surface-potential-based compact model for GaN HEMT IC design," *IEEE Transactions on Electron Devices*, **60**(10), 3216-3222 (2013).

Perimeter Gated Single Photon Avalanche Diodes in Sub-Micron and Deep-Submicron CMOS Processes

Mst Shamim Ara Shawkat*, Mohammad Habib Ullah Habib, Md Sakib Hasan,
Mohammad Aminul Haque and Nicole McFarlane

*Department of Electrical Engineering and Computer Science,
University of Tennessee, Knoxville, TN 37996, USA*
mshawkat@vols.utk.edu

A perimeter gated SPAD (PGSPAD), a SPAD with an additional gate terminal, prevents premature perimeter breakdown in standard CMOS SPADs. At the same time, a PGSPAD takes advantage of the benefits of standard CMOS. This includes low cost and high electronics integration capability. In this work, we simulate the effect of the applied voltage at the perimeter gate to develop a consistent electric field distribution at the junction through physical device simulation. Additionally, the effect of the shape of the device on the electric field distribution has been examined using device simulation. Simulations show circular shape devices provide a more uniform electric field distribution at the junction compared to that of rectangular and octagonal devices. We fabricated PGSPAD devices in a sub-micron process (0.5 μm CMOS process and 0.5 μm high voltage CMOS process) and a deep-submicron process (180 nm CMOS process). Experimental results show that the breakdown voltage increases with gate voltage. The breakdown voltage increases by approximately 1.5 V and 2.5 V with increasing applied gate voltage magnitude from 0 V to 6 V for devices fabricated in 0.5 μm and 180 nm standard CMOS process respectively.

Keywords: Single photon avalanche diode (SPAD); perimeter gated single photon avalanche diode (PGSPAD); CMOS.

1. Introduction

Single photon avalanche diodes (SPADs) have gained interest for use in a wide range of applications such as biochemical analysis, imaging and light ranging applications [1–9]. A SPAD is a p-n junction operating in Geiger mode, i.e. the device is biased above the breakdown voltage. When a single photon is incident on the device, the SPAD generates a large avalanche current through impact ionization. Commercial SPADs are typically fabricated in customized optical CMOS process to allow for control of the avalanche process [5]. Due to the planar nature of the junction, the diode periphery undergoes breakdown earlier compared to other parts of the junction, this is known as premature breakdown [1–9]. Premature breakdown is one of the main problems of SPADs implemented in standard CMOS process and adversely affects the performance. Perimeter gated SPAD (PGSPAD), a SPAD with an additional gate terminal has been proven to prevent premature breakdown in standard CMOS process while keeping the benefits of low cost and electronic integration [5], [7], [10], [11], [12]. The breakdown voltage is modulated in PGSPADs by applying a voltage at the perimeter gate. These PGSPADs have been implemented in submicron CMOS process [5], [7], [10–13].

Additional challenges arises in designing avalanche photodiodes in the deep-submicron CMOS processes than in larger processes. In order to create a large reasonable active area and operate in the Geiger mode in the deep submicron CMOS processes, it is required to design the SPAD in a configuration that supports a planar and uniform multiplication region that expands both laterally and vertically [14]. Additionally, the noise due to the presence of shallow-trench isolation (STI), reduced annealing, and drive-in diffusion steps and higher doping levels are major issues in deep submicron CMOS SPADs. With increased doping levels, noise due to tunneling and parasitic capacitance increases. Since the avalanche discharge increases with increasing parasitic capacitance, this degrades the afterpulsing probability. The concentration of impurities is higher due to reduced annealing steps results in increased thermally generated noise [15]. Therefore, it is necessary to study the effect of perimeter-gated technique in the deep submicron CMOS processes. In this work, we present and verify the effectiveness of perimeter gated technique of PGSPADs fabricated both in sub-micron and deep-submicron CMOS processes through experiment. The effectiveness of the applied voltage at the perimeter gate to generate a uniform electric field distribution at the junction has been verified through the TCAD device simulations. Effect of device shape on the electric field distribution has been examined through physical device simulation.

2. Working principle of SPAD

A single photon avalanche diode (SPAD) is a p-n junction, which is biased above the breakdown voltage. The device operates in the so-called Geiger mode. With the increase of reverse bias voltage beyond a critical potential, the current increases rapidly. The rapid increase in current is due to the free carriers, which have gained enough energy to ionize the fixed lattice atoms. The electric fields in the junction are very strong due to the applied reverse bias. As a result, the newly freed carriers are accelerated with large energy giving rise and can free other carriers through high-energy collisions. This rapid multiplication process results in a sudden large avalanche current. The reverse voltage above which this multiplication process occurs is called breakdown voltage [1–9].

For a SPAD to respond to an incident photon, impact ionization is necessary. A high electric field is required to break the covalent bonds and generate the free carriers, which take part in the avalanche process. Since the electric field is proportional to the breakdown voltage, higher breakdown voltages are needed for impact ionization. The number of electron-hole pairs produced by a free carrier per unit length traveled determines the ionization rate, α [16]. The ionization rate of electrons and holes are different due to the differing effective mass. The ionizing carrier has to acquire enough energy, which is at least a threshold energy from the electric field. Assuming both electrons and holes have equal effective mass, the minimum required energy is $1.5 \times Eg$ based on the laws of momentum at a collision event and conservation of energy for impact ionization [16]. The probability that carriers gain the threshold energy is a function of the local electric field and previous states of the carrier. This probability determines the ionization rates. Figure 1 illustrates impact ionization causing an avalanche current in the presence of a high electric field. A number of free carriers are generated at each step, quickly multiplying and generating a large current.

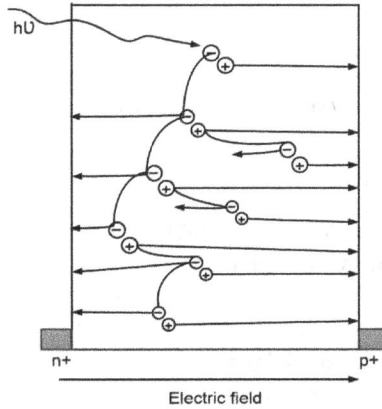

Fig. 1. Impact ionization causing avalanche current in existence of high electric field where number of free carriers are multiplied in each step.

The empirical expression for local avalanche generation is [16–17],

$$\alpha_{n,p} = a_{n,p} \exp\left(-\frac{b_{n,p}}{E}\right) \tag{1}$$

where a and b are the ionization coefficients, and E represents the electric field in the direction of current. For holes and electrons, the value of a and b are different.

In SPADs, there is no current when there are no free carriers. When a photon arrives, free carriers are generated resulting in an avalanche current, which increases rapidly. Free carriers are generated in absence of photon due to inherent noise processes such as thermal generation, minority carrier diffusion, and band-to-band tunneling. The large avalanche current must be quenched to protect the long term use of the device. An external circuit, typically a passive ballast resistor or a transistor, is employed to quench the large avalanche current. As the current increases the voltage drop across the quenching resistor increases. As a result, the bias voltage across the SPAD reduces below the breakdown voltage ending the avalanche process. The SPAD can is then recharged and ready to detect photon event induced avalanche. The avalanche gain is large enough to be almost infinite in this Geiger mode of operation. The generation of an electron-hole pairs causes the device to be in an "ON" or "OFF" state. Since the generation of an avalanche gives a current or voltage spike, the device exhibits an inherently digital operation [1–9]. Figure 2 shows the Geiger mode operation of SPAD and quenching of the avalanche current.

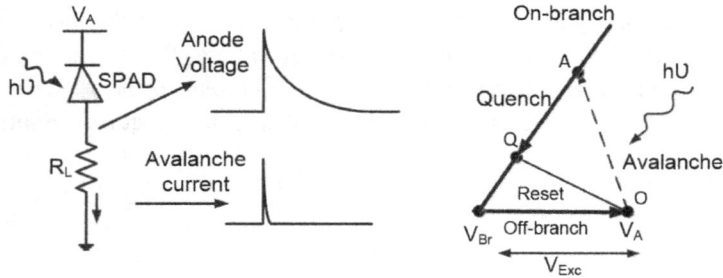

Fig. 2. Geiger mode operation of SPAD. The avalanche current is quenched using a quenching resistor.

3. Limitations of CMOS SPADs

In general, dedicated processes are used to fabricate SPADs due to their well-controlled avalanche profiles. However, this can preclude the use of electronics integrated into these commercial processes [5, 9]. However, there is increasing interest in porting SPAD fabrication to mainstream CMOS processes due to the benefits of standard CMOS process [1–9], [15]. One of the most beneficial features of CMOS technology is the ability to integrate SPADs with a wide variety of electronics to increase functionality. This includes integrating quenching, sensing electronics and digital processing blocks into a single chip. In addition, standard CMOS process provides other benefits such as low-cost fabrication and relatively low dark current (noise) [1–9], [16]. However, due to the planar nature of the junction in standard CMOS process, electric field distributions exhibit maxima at the diodes' edge [17]. Since SPADs operate in a high reverse bias region, they are more vulnerable to breakdown at the edges. Therefore, full volumetric breakdown cannot be achieved as the diode periphery undergoes breakdown earlier compared to the other parts of the diode. As a result, the active area is decreased and the photon detection efficiency is significantly compromised. This premature breakdown is one of the major problems of SPADs implemented in standard CMOS process [1–9].

Several methods have been reported to prevent premature breakdown in CMOS avalanche diodes. A guard ring has been used to prevent the generation of a high electric field at the p⁺ anode edge for a photodiode operated in avalanche mode and fabricated in a standard CMOS process. In [18], the advantage of lateral diffusion of two n-wells maintaining optimal distance was explored to create the guard ring without any further processing steps. A lighter n-doped region at the periphery of the p-n junction was created using lateral diffusion of donor atoms following the n-well oxidation (Fig. 3(a)).

In [19], a field limiting guard ring is placed at a minimum distance from the p⁺ implant in combination of a control gate over the gap to reduce the premature breakdown. The designed photodiode is fabricated in a CMOS compatible process in [19]. However, the fill factor is reduced with the use of guard ring and is therefore not always an ideal option [19]. Dandin *et al.* have shown that the combination of lateral diffusion of n-wells with a depletion gate lessens the premature edge breakdown [21] (Fig. 3(b)). All of these techniques are based on the effect of the modulation of the dopant concentration and

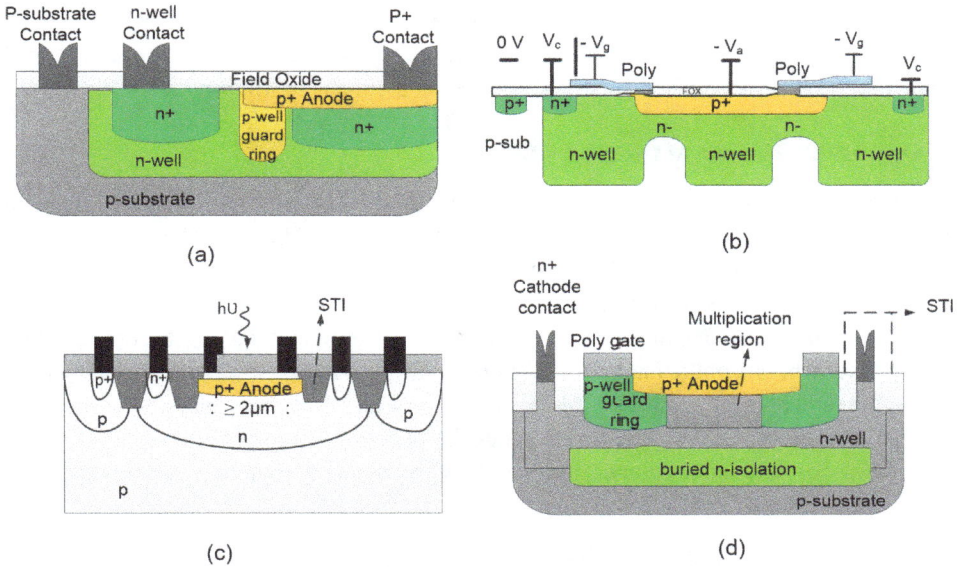

Fig. 3. Different methods to prevent the premature edge breakdown in CMOS SPADs: (a) The anode is surrounded with a guard-ring to prevent premature breakdown [18]. (b) lateral diffusion of two adjacent n-wells creates a lighter doped region at the edges of the junction and a poly control gate further depletes the surface to prevent premature breakdown [21], (c) The silicon-dioxide trenches confines the p+ drain implant, thereby ensuring a planar junction and a uniform avalanche breakdown [23], and (d) A guard ring of p-well around the p+ anode within n-well cathode is used to prevent premature breakdown [15].

junction curvature on the breakdown voltage as well as the effect of the gate on top of the high field regions [20, 22].

Designing SPADs in deep submicron CMOS processes faces additional challenges compared to that of larger processes. Deep-submicron CMOS process use shallow trench isolation (STI). A planar and uniform multiplication region expanding both laterally and vertically under the area of SPAD is required to maintain Geiger mode operation of SPADs in deep submicron [14]. The noise performance of the SPADs implemented in deep sub-micron CMOS process is reduced due to the higher doping levels, the presence of STI, reduced annealing, and reduced drive-in diffusion steps [15]. As the doping levels increase, the noise due to tunneling and parasitic capacitances increases. With the increase of the parasitic capacitance, the afterpulsing probability worsens through increasing the avalanche discharge. Several methods have been reported in literature to mitigate the challenges of SPADs in deep sub-micron CMOS process. In [23], the junction's geometry has been modified where STI has been used advantageously to prevent the premature edge breakdown (Fig. 3(c)). Niclass *et al.* has used a guard ring of p-well around the p+ anode to prevent the premature breakdown [15]. The fabricated SPAD was octagonal shape and used a p+/n-well junction with a guard ring around the anode shown in Fig. 3(d) [15].

The solutions for deep submicron processes tend to increase the size of the device. Placement of a perimeter gate on top of the junction and application of the voltage on that gate are also effective ways of preventing premature breakdown while having a large fill factor [5, 7, 21–24]. The placement of the perimeter gate results in the perimeter gated single photon avalanche diode (PGSPAD) which is the focus of this work.

4. Operating principle of CMOS PGSPAD

Perimeter gated single-photon avalanche diode (PGSPAD) is a p-n junction with an additional poly-silicon gate surrounding the junction [5], [7], [12–15]. The layout and the cross sectional view of PGSPAD device is presented in Fig. 4. The applied voltage at the gate terminal modulates the electric field making it uniform throughout the junction. The perimeter gating technique has been proven to be an efficient method to prevent premature edge breakdown, one of the major problems in operating planar p-n junctions of avalanche diodes implemented in standard CMOS process during avalanche breakdown [5], [7], [10–13]. Therefore, CMOS PGSPADs adopt the benefits of commercial CMOS process such as low cost and integration capability while preventing premature edge breakdown.

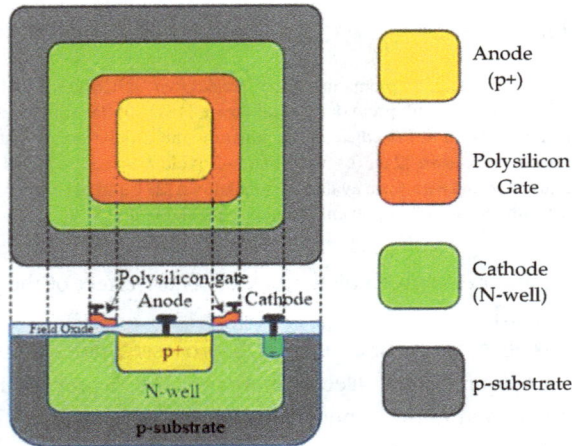

Fig. 4. Layout and cross-sectional view of PGSPAD device. Additional gate bias is applied at the additional polysilicon gate terminal of PGSPAD to prevent premature edge breakdown through making uniform electric field at the junction.

In order to operate in Geiger mode, the PGSPAD device is biased beyond the breakdown voltage similar to that of regular SPAD. The charge carriers are freed by the incoming photon. The free carriers are then accelerated by the high electric field generated by the high reverse bias voltage. These carriers go through the impact ionization in the depletion region producing a self-sustaining avalanche of carriers [25–26]. The avalanche gain or multiplication factor, M, is [26],

$$M = \left\{ 1 - \int_0^{W_D} \alpha_n \exp\left[- \int_x^{W_D} (\alpha_n - \alpha_p) dx \right] dx \right\}^{-1} \qquad (2)$$

where , W_D is the depletion-layer width, and α_n and α_p are the ionization rates for electrons and the holes respectively [26]. Assuming equal ionization rates i.e. $\alpha_n = \alpha_p = \alpha$, the multiplication factor becomes,

$$M = \frac{1}{1 - \alpha W_D} \tag{3}$$

Breakdown occurs when $\alpha W_D = 1$ [26].

For a regular SPAD, when the electric field is applied across the device, the electric field is maximum at the sharp edges causing premature breakdown. In PGSPADs, the premature breakdown can be suppressed through the applied voltage at the additional gate terminal through modulation of carrier concentration [5]. The breakdown voltage, V_{Br}, a parameter that is correlated with efficient SPAD operation, is [27],

$$V_{Br} = \frac{\varepsilon_s E_{crit}^2}{2eN_b} \tag{4}$$

where V_{Br} is the breakdown voltage, E_{crit} is the critical electrical field at breakdown, N_b is the doping concentration of the highly doped region, e is the charge of an electron, and ε_s is the permittivity. In [20], it was experimentally verified that the breakdown voltage can be modulated using an external field at the surface. Either side of the junction can be depleted through the variation of the applied external field. Therefore, the surface field tends to saturate at a maximum and at a minimum value with the variation of gate voltage. The induced field in the junction places an upper and lower limit of the breakdown voltage variation preventing further variation in shape of the depletion region. The breakdown voltage is defined between these two extremes in [21]

$$V_{Br} = mV_G + \beta \tag{5}$$

where V_G is the applied gate voltage, β is a constant, and m is the slope which approaches unity for small oxide thickness and low substrate impurity concentrations.

Therefore, the breakdown voltage can be changed in PGSPADs by applying the voltage at the gate based on the empirical expression. Since the performance of the device depends on the breakdown voltage, the performance can be improved through applied gate voltage [10–13]. Moreover, premature edge breakdown prevention was shown to help in minimizing the dark count rate of a PGSPAD resulting in photon counting with high and tunable signal-to-noise (SNR) ratios [12].

5. Mathematical formulation of PGSPAD operation

In a p-n junction under a high enough reverse potential, charge carriers with high velocities can undergo scattering and collide with electrons in the valence band. If it can transfer energy larger than bandgap energy to an electron upon collision, the electron move to the conduction band creating a free electron-hole pair for conduction. The generation rate of electrons and holes in the semiconductor for specific carrier concentrations, n and p is,

$$\frac{dn}{dt} = \frac{dp}{dt} = \alpha_n n v_e + \alpha_p p v_h \qquad (6)$$

where v_e and v_h are the velocities of the charge carriers. α_n and α_p are the ionization rates of electrons and holes (i.e. generated electron-hole pairs per unit distance) [26]. Since the velocity of a carrier is determined by the strength of the electric field, ionization rates are also dependent on electric field and is given as [28],

$$\alpha = \frac{qE}{E_i} \exp\left(-\frac{E_i}{E\left(1 + \frac{E}{E_\tau}\right) + E_{kT}}\right) \qquad (7)$$

where, E_i is the threshold for high electric field required for ionization and E_{kT}, E_τ, and E_i are threshold electric fields to overcome deceleration caused by thermal, optical-photon and scattering effects through changing carrier dynamics. The rate of ionization (electrons or holes) in silicon can be related to the multiplication factor [29],

$$1 - \frac{1}{M} = \int_0^W \alpha_i dx \qquad (8)$$

where, M ($=n/n_o$) is the multiplication factor, n_o is number of electrons injected into the junction, and α_i is rate of ionization. When this integral becomes 1, the multiplication factor becomes infinite, resulting in an avalanche of carriers and ultimately causing avalanche breakdown. The multiplication factor depends entirely on the electric field. This is consistent with the assumption that carriers lose energy through collisions with lattice atoms rather than collisions to create ionization. The time required for a carrier under high enough electric field (to generate multiplication) to traverse the junction is on the order of 10^{-10}s. This is negligible compared to the time for recombination ($>10^{-6}$s) and as such carrier recombination is typically ignored.

In conventional CMOS process, doping is graded (non-uniform). For a linearly graded p-n junction (where N_D and N_A are donor and acceptor concentrations), $N_D - N_A = ax$ and a is a material and doping specific constant. The value of a determines the rate at which the space charge density and electric field changes while going across the junction. The electric field is parallel to the space coordinate x, and $x = 0$ corresponds to the center of the junction. The relationship among electric field E, junction voltage V, and junction width W in a linearly graded junction are,

$$E = E_M\left[1 - \left(\frac{2x}{W}\right)^2\right] \qquad (9)$$

$$E_M = \frac{1.5 \times V}{W} \tag{10}$$

$$W = W_1 V^{1/3} = \sqrt{\frac{2}{3} \times W_1^3 E_M} \tag{11}$$

$$And \quad V = V_a + V_{bi} \tag{12}$$

where, E_M is maximum electric field across the junction, V_a is applied voltage, $V_{bi}=$ built-in voltage (≈ 0.7 V for Si at room temperature), and W_1 is a constant.

The device under consideration in this article is perimeter gated SPAD. To enable photon detection, the PGSPAD is not operated in forward biased mode. Rather it is operated in avalanche mode under reverse biased condition. While in reverse bias, a reverse saturation current flows through the diode. The reverse current density is determined by multiple factors, current density developed by space charge within the space charge region J_{scg}, hole current density J_{po}, and electron current density J_{no}. Assuming a constant thermal generation rate for electrons and holes within the space charge region, the current density is [30],

$$J = M_p J_{po} + M_n J_{no} + M_{scg} J_{scg} \tag{13}$$

where M_p, M_n, and M_{scg} are the avalanche multiplication coefficients expressed as,

$$M_n = \frac{1}{1 - exp[-\psi(W)]\left\{\int_0^W \alpha_n(x)exp[\psi(x)]dx\right\}} \tag{14}$$

$$M_p = exp[-\psi(W)]M_n \tag{15}$$

$$M_{scg} = \left\{\frac{1}{W}\int_0^W exp[\psi(x)]dx\right\}M_n \tag{16}$$

$$\psi(x) = \int_0^x [\alpha_n(\omega) - \alpha_p(\omega)]d\omega \tag{17}$$

where, α_n is the electron ionization rate and α_p is the hole ionization rate. Avalanche breakdown takes place when these coefficients get close to infinity, resulting in the generic breakdown criterion,

$$exp[-\psi(W')]\left\{\int_0^{w'} \alpha_n(x)exp[\psi(x)]dx\right\} = 1 \tag{18}$$

where W' is the depletion region width during avalanche breakdown. When $\alpha_n = \alpha_p$ and are independent of electric field, the breakdown criterion reduces to,

$$\int_0^{W'} \alpha_n(x)dx = 1 \tag{19}$$

As mentioned before, planar p-n junctions tend to experience high electric field at the junction edges. This leads to premature breakdown of the junction. This is due to the presence of charges on the surface causing the electric field lines to terminate at these surface charges rather than on the immobile ion cores in the space charge region [31]. This changes the shape and intensity profile of electric field thereby changing the breakdown voltage. This change in breakdown voltage as an effect of gate voltage variation can be approximated by Eq. (5) as discussed in the previous section. The electric field at the corner of the pn junction can be expressed using,

$$E_{corner} = \frac{V_R}{W_f} + \frac{V_R - V_G}{3C_{ox}} \tag{20}$$

where, E_{corner} is electric field at the pn junction periphery near the surface, W_f is space charge region width far from surface where the change in potential function between gate and silicon within the insulator is linear, V_R is cathode potential of the p-n junction diode, C_{ox} is oxide thickness under the gate. The relationship between the breakdown voltage and gate voltage is,

$$V_{Br} = \frac{1}{1 + \frac{3C_{ox}}{W_f}} V_G + \frac{3E_{crit}C_{ox}}{1 + \frac{3C_{ox}}{W_f}} \tag{21}$$

where E_{crit} is the critical electric field at breakdown. There is a linear relationship between the breakdown voltage and the gate voltage. This also shows the dependency of the breakdown voltage on the oxide thickness which is a process dependent parameter [32]. For instance, one popular process is publicly available 0.5 μm CMOS process where oxide thickness, $t_{ox} = 1.41 \times 10^{-8}$ cm and $E_{crit} \approx 2.36 \times 10^5 \, V/cm$ at $V_R = 14 \, V$ for the doping concentrations used in the following section.

6. Effect of gate voltage on modulation of electric field distribution

In this work, the effectiveness of modulation of electric field through the applied gate voltage has been verified using Sentaurus device simulation. The electric field distribution for a regular SPAD is also simulated for performance comparison. The implemented junction is a p$^+$/n-well diode embedded in p-type bulk substrate. For the simulation, the doping profile of [5] was used and is repeated here for completeness. For the p$^+$ region the peak concentration is 1×10^{20} cm^{-3} while the peak concentration for the n-well used is 1.22×10^{17} cm^{-3} with Gaussian profile. The concentration reduces with $\sigma = 120$ nm in the lateral direction and with $\sigma = 50$ nm in the depth direction. A shifted Gaussian profile was used with a mean of 0.2 μm from the surface. For the p-substrate, the doping concentration of 1×10^{15} cm^{-3} is used.

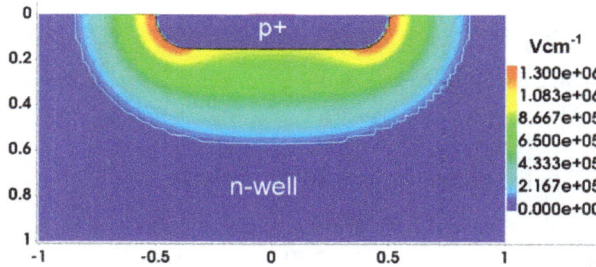

Fig. 5. Simulation result of electrical field modulation of a regular SPAD (x-axis and y-axis are in μm).

Simulation results presented in Fig. 5 show the electric field distribution for a regular SPAD without any gate terminal. Electric field shows maxima at the diode's edge. Therefore diode edge undergoes avalanche earlier compared to the other part of the junction resulting premature breakdown, one of the major problem of SPAD in CMOS process.

The electric field distribution for a PGSPAD with different applied gate voltages are presented in Fig. 6 and Fig. 7. Figure 6 shows as the applied voltage magnitude at the additional gate terminal of PGSPAD increases from 0 V to 4 V, the electric field decreases at the edges and starts to become uniform throughout the junction compared to those at regular SPAD. Figure 7 shows if the applied gate voltage increases further to 8 V, the electric fields become more uniform and with 12 V the electric field distribution around the junction becomes almost fully uniform. This is due to the modulation of carrier concentration at the junction through the applied gate voltage. Therefore, the premature edge breakdown is prevented and full volumetric breakdown is achieved.

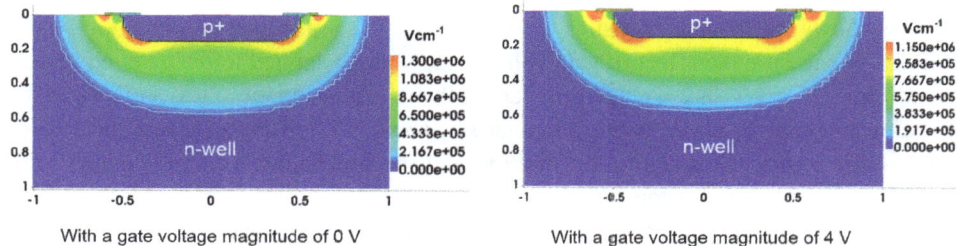

With a gate voltage magnitude of 0 V   With a gate voltage magnitude of 4 V

Fig. 6. Simulation result of electrical field modulation using applied gate voltage magnitude of 0 V and 4 V (x-axis and y-axis are in μm).

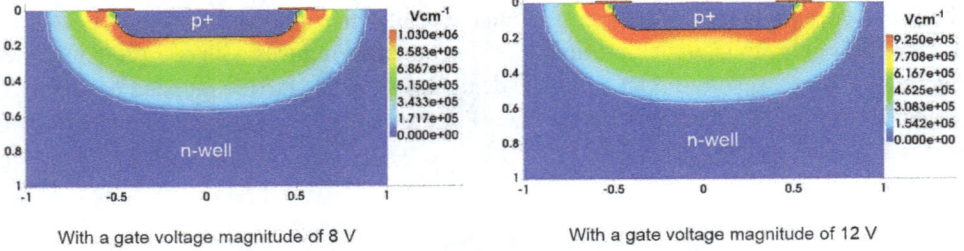

With a gate voltage magnitude of 8 V   With a gate voltage magnitude of 12 V

Fig. 7. Simulation result of electrical field modulation using applied gate voltage magnitude of 8 V and 12 V (x-axis and y-axis are in μm).

7. Effect of applied gate voltage on breakdown voltage

The experimental results reported here are the average values of three tested chips with a maximum deviation of approximately 2.81%. Photomicrographs of the fabricated PGSPAD device in sub-micron standard CMOS processes, 0.5 μm CMOS process, and 0.5 μm high voltage CMOS process, are shown in Fig. 8. In order to verify the effectiveness of the additional gate terminal of PGSPAD device, we fabricated similar PGSPAD device in sub-micron processes, 0.5 μm CMOS process, 0.5 μm high voltage CMOS process, and deep-submicron process, 180 nm CMOS process, with size of 22 μm × 22 μm.

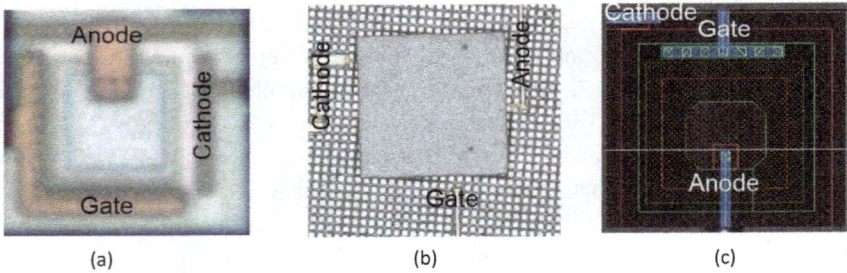

(a)   (b)   (c)

Fig. 8. Photomicrographs of the fabricated PGSPAD device in standard CMOS process: (a) 0.5 μm CMOS process, (b) 0.5 μm high voltage CMOS process, and (c) 180 nm CMOS process.

Figure 9 shows the IV characteristics of PGSPAD device implemented in 0.5 μm CMOS process. To measure the effect of gate voltage, the voltage applied to the gate terminal is swept from 0 V to 8 V in 2 V steps while doing the IV characteristics. It is shown that the breakdown voltage increases with gate voltage by about 1.5 V. This is because the applied gate voltage varies the carrier concentration and the breakdown voltage changes with change of carrier concentration.

The modulation of breakdown voltage by the applied gate voltage is also verified in 0.5 μm high voltage CMOS process. Figure 10 shows the IV characteristics of PGSPAD device implemented in 0.5 μm high voltage CMOS process. The applied gate voltage is swept from 0 V to 8 V in 2 V steps while doing the IV characteristics. It is shown that the breakdown voltage increases by around 1.5 V in a similar way with gate voltage. This is because the applied gate voltage varies the breakdown voltage through changes of carrier concentration.

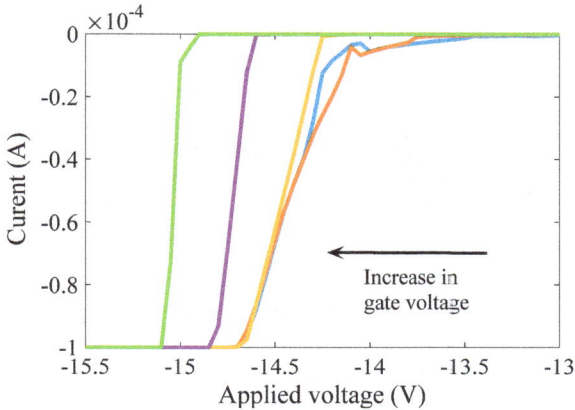

Fig. 9. IV characteristics of PGSPAD device implemented in 0.5 μm CMOS process.

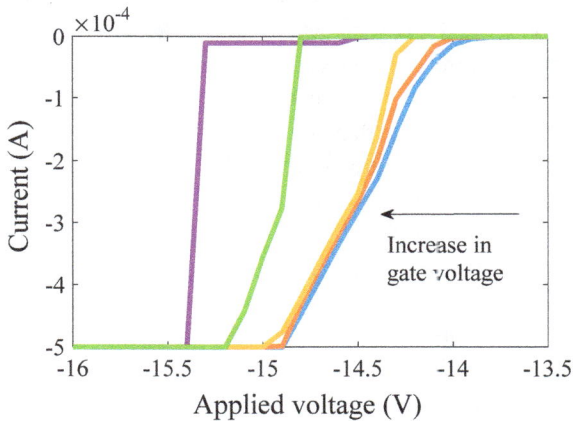

Fig. 10. IV characteristics of PGSPAD device implemented in 0.5 μm high voltage CMOS process.

Figure 11 shows the IV characteristics of PGSPAD device implemented in 180 nm CMOS process. For the deep-submicron process, the breakdown voltage also increases with gate voltage by about 2.5 V as expected. This is due to the modulation of the carrier concentration with applied gate voltage and the change of breakdown voltage with change of carrier concentration. Therefore, PGSPAD fabricated in lower process also shows the similar trend in breakdown voltage as in larger process.

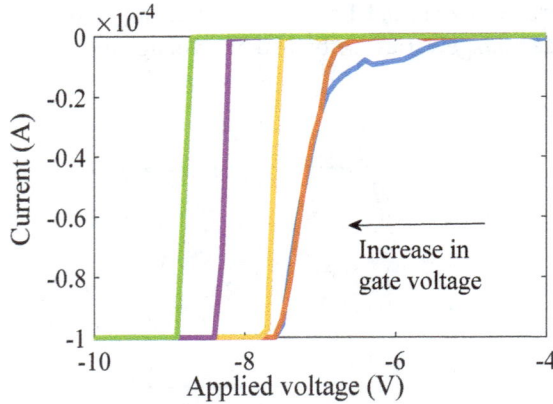

Fig. 11. IV characteristics of PGSPAD device implemented in 180 nm CMOS process.

8. Effect of shape on electric field distribution

The effect of shape on the electric field distribution has also been verified through the device simulation. The electric field is not uniformly distributed around the junction of the device. For this reason, the whole junction does not enter breakdown at the same time. The device will breakdown in the regions where the breakdown field is reached first. The electric field at the corners of the device reaches the breakdown field before other regions. The chances of edge breakdown can be reduced by changing the shape of the device.

Results presented in Fig. 12 show that edge breakdown can be reduced from changing the shape of the device from square into circular. This is because the square shape device has more sharp edge causing electric field maxima at the edges results in premature edge breakdown compared to that of circular shape device. Since the octagonal shape device has less sharp edge compared to that of square shape, it shows more uniform electric field distribution compared to that of square shape device (Fig. 12(a) and (b)). The circular shape device has the most uniform electric field distribution at the junction, as its edges are not sharp like that of the square and octagonal shape device.

We fabricated square, octagonal, and circular devices to observe the variation in the breakdown voltage as a function of the applied gate voltage. We simulated devices with different shapes using Sentaurus device simulator (Fig. 13) and observed variations in breakdown voltage (Fig. 14). The devices simulated were p-sub/n-well devices.

Our simulations show the breakdown voltage for the three different shapes and confirms our experimental measurements and theoretical considerations that circular devices have higher breakdown voltages than octagonal which have higher breakdowns than square devices at zero voltage applied to the gate of the device.

As observed from Table 1 and Fig. 13, for the small devices with the same junction type (p^+/n-well), the breakdown voltage for the circular devices are higher than that of the octagonal or the square shaped junctions for any values of the gate voltage from 0V to -8V. However, the octagonal junction has higher breakdown voltage when the gate voltage is 0V. But, for the gate voltages beyond -0.4V, the breakdown voltage for the square junction is higher than the device with octagonal junction. Increasing the corner angles reduces the electric field and a change is breakdown voltage is observed [25].

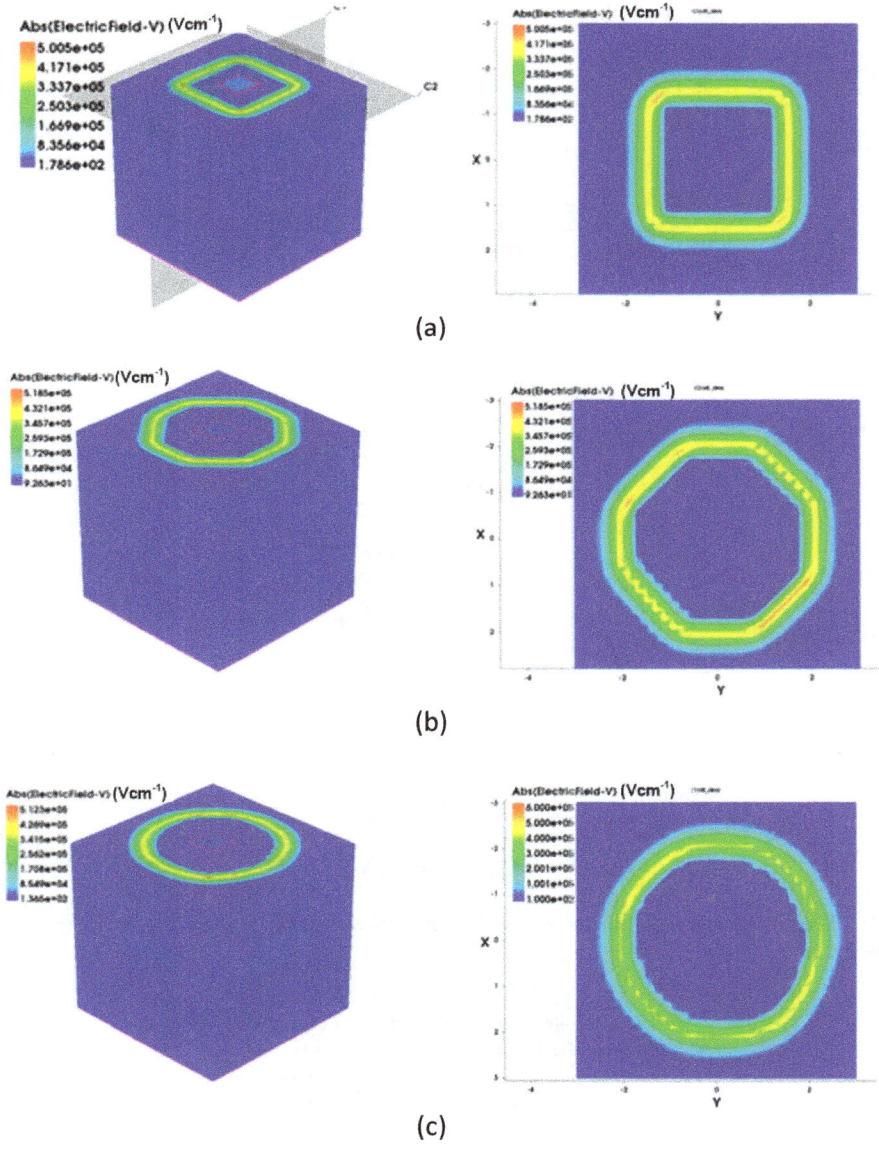

(a)

(b)

(c)

Fig. 12. Electric field distribution simulation for PGSPADs with different shapes: (a) square shape, (b) octagonal shape, (c) circular shape (x-axis and y-axis are in μm).

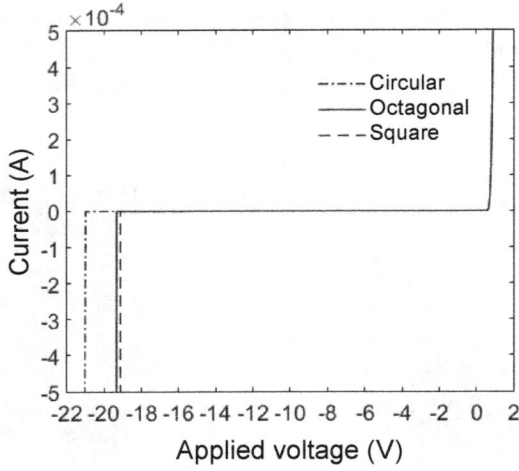

Fig. 13. Simulated IV characteristics showing the change in breakdown voltage for different shapes (p-sub/n-well) at zero gate voltage.

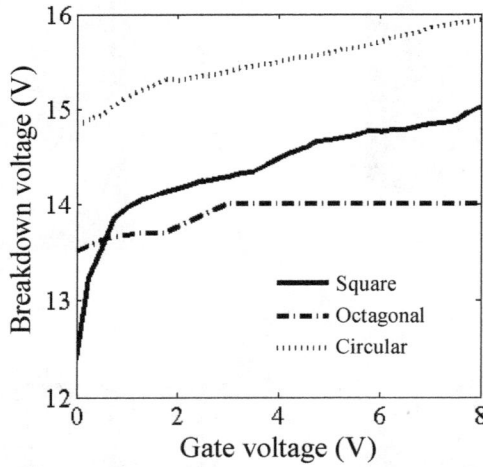

Fig. 14. Breakdown voltage vs applied gate voltage for different shapes (square, circular [33]) and small p$^+$/n-well junction.

Table 1. Fabricated small p$^+$/n-well devices with different shapes (v_{Br} = breakdown voltage).

Diode Type	V_{Br} (V)
square p$^+$/n-well	12.4
octagonal p$^+$/n-well	13.5
circular p$^+$/n-well	14.8

9. Conclusion

This paper reviews the effectiveness of the perimeter-gated technique used in PGSPAD devices fabricated in varying standard CMOS processes. The effectiveness of the applied voltage at the perimeter gate to make a uniform electric field at the junction preventing the premature breakdown was simulated through Sentaurus device simulation. Simulation results show that increasing the applied gate voltage magnitude makes the electric field around the junction more uniform, preventing the premature edge breakdown. The effect of the shape of the device on the electric field distribution was also simulated. The probability of premature edge breakdown can be reduced by changing the shape of the device. From device simulation results, the circular shaped device provides a more uniform electric field distribution at the junction compared to that of the rectangular and octagonal devices. Experimental results verified that the breakdown voltage changes in PGSPADs using the voltage at the gate. Since the device performance is correlated with breakdown voltage, the performance of the device can be improved using PGSPADs. PGSPAD devices fabricated in sub-half micron processes also show a similar trend in breakdown voltage as in the larger processes providing similar improvement.

Acknowledgement

This material is based upon work supported by the U.S. Department of Energy, Office of Science, Office of Basic Energy Sciences under Award Number DE-SC0017414.

References

1. S. Cova, M. Ghioni, A. Lotito, I. Rech, and F. Zappa, "Evolution and prospects for single-photon avalanche diodes and quenching circuits," *Journal of Modern Optics*, vol. 51, pp. 1267-1288, Jun. 2004.
2. C. Niclass, A. Rochas, P. Besse, and E. Charbon, "Toward a 3-D camera based on single-photon avalanche diodes," *IEEE Journal of selected topics in Quantum Electronics*, vol. 10, pp. 796-802, Jul. 2004.
3. C. Niclass, A. Rochas, P. Besse, and E. Charbon, "Design and characterization of a CMOS 3-D image sensor based on single-photon avalanche diodes," *IEEE Journal of Solid-State Circuits*, vol. 40, pp 1847-1854, Sep. 2005.
4. M. Gersbach, Y. Maruyama, C. Niclass, K. Sawada, and E. Charbon, "A room temperature CMOS single photon sensor for chemiluminescence detection," *Miniaturized Systems for Chemistry and Life Sciences* , Tokyo, Japan, Nov. 2006.
5. M. Dandin, A. Akturk, B. Nouri, N. Goldsman, and P. Abshire, "Characterization of single-photon avalanche diodes in a 0.5μm standard CMOS process - Part 1: perimeter breakdown suppression," *IEEE Sensors Journal,* vol. 10, no. 11, pp. 1682-1690, Nov. 2010.
6. D. Stoppa, D. Mosconi, L. Pancheri, and L. Gonzo, "Single-photon avalanche diode CMOS sensor for time-resolved fluorescence measurements," *IEEE Sensors Journal*, vol. 9, no. 9, pp. 1084-1090, Sep. 2009.

7. M. Dandin, M. H. U. Habib, B. Nouri, P. Abshire and N. McFarlane, "Characterization of single-photon avalanche diodes in a 0.5μm standard CMOS process - Part 2: equivalent circuit model and Geiger mode readout," *IEEE Sensors Journal*, vol. 16, no. 9, pp. 3075-3083, May, 2016.

8. D. P. Palubiak and M. J. Deen, "CMOS SPADs: Design issues and research challenges for detectors, circuits, and arrays," *IEEE Journal of Selected Topics in Quantum Electronics*, vol. 20, no. 6, pp. 409-426, Nov. 2014.

9. B. Aull, A. Loomis, D. Young, D. J. Young, R. M. Heinrichs, B. J. Felton, P. J. Daniels, and D. J. Landers, "Geiger-mode avalanche photodiodes for three dimensional imaging," *Lincoln Laboratory Journal*, vol. 13, pp. 335-345, Nov. 2002.

10. M. H. U. Habib, F. Quaiyum, S. Islam, and N. McFarlane, "Optimization of perimeter gated SPADs in a standard CMOS process," *IEEE Sensors Conference*, pp. 1668-1671, Nov. 2014.

11. M. H. U. Habib and N. McFarlane, "A tunable dynamic range digital single photon avalanche diode," *IEEE Electron Device Letters*, vol. 38, no. 1, pp. 60-63, Jan. 2017.

12. M. S. A. Shawkat and N. McFarlane, "A CMOS perimeter gated SPAD based mini-digital silicon photomultiplier," *IEEE International Midwest Symposium on Circuits and Systems*, Windsor, Canada, 4 pages, Aug. 2018.

13. M. H. U. Habib, M. S. A. Shawkat, and N. McFarlane, "A tunable single photon avalanche diode pixel with improved time resolution," *IEEE Sensors Conference*, pp. 1-3, Oct. 2017.

14. S. Cova, M. Ghioni, F. Zappa, I. Rech., and A. Gulinatti, " A view on progress of silicon single photon avalanche diodes," *SPIE Advanced Photon Counting Techniques*, vol. 6372, p. 63720I-1, Oct. 2006.

15. C. Niclass, M. Gersbach, R. Henderson, L. Grant, and E. Charbon, "A Single Photon Avalanche Diode Implemented in 130-nm Technology, " *IEEE Journal of Selected Topics in Quantum Electronics*, vol. 13, pp. 863-869, Jul. 2007.

16. W. Maes, K. D. Meyer, and R. V. Overstraeten, "Impact ionization in silicon: A review and update," *Solid-State Electronics*, vol. 33, no. 6, pp. 705-718, Jun. 1990.

17. A. G. Chynoweth, "Ionization rates for electrons and holes in silicon," *Physical Review*, vol. 109, no. 5, pp. 1537-1540, Mar. 1958.

18. A. Rochas, A. R. Pauchard, P.-A. Besse, D. Pantic, Z. Prijic, and R. S. Popovic, "Low-noise silicon avalanche photodiodes fabricated in conventional CMOS technologies," *IEEE Transactions on Electron Devices*, vol. 49, pp. 387-394, Mar. 2002.

19. A. Pauchard, P.A. Besse, and R. S. Popovic, "Simulation of a new CMOS compatible method to enhance the breakdown voltage of highly doped shallow PN junctions," *International Conference Modeling and Simulation of Microsystems*, pp. 420-425, 1998.

20. V. A. K. Temple and M. S. Adler, "Calculation of the diffusion curvature related avalanche breakdown in high-voltage planar p-n junctions," *IEEE Transactions on Electron Devices*, vol. 22, pp. 910-916, Oct. 1975.

21. M. Dandin, N. Nelson, V. Saveliev, H. Ji, P. Abshire, and I. Weinberg, "Single photon avalanche detectors in standard CMOS," *IEEE Sensors Conference*, pp. 585-588, Oct. 2007.

22. A. S. Grove, O. Leistiko, Jr., and W. W. Hooper, "Effect of surface fields on the breakdown voltage of planar silicon p-n junctions," *IEEE Transactions on Electron Devices*, vol. 14, pp. 157-162, Mar. 1967.

23. H. Finkelstein, M. Hsu, and S. Esener, "STI-bounded single-photon avalanche diode in a deep submicrometer CMOS technology," *IEEE Electron Device Letter*, vol. 27, pp. 887-889, Nov. 2006.

24. M. Dandin and P. Abshire, "High Signal-to-Noise Ratio Avalanche Photodiodes With Perimeter Field Gate and Active Readout," *IEEE Electron Device Letters*, vol. 33, no. 4, pp. 570-572, Apr. 2012.

25. M. W. Fishburn, "Fundamentals of CMOS Single-Photon Avalanche Diodes", Fishburn, 2012.

26. S. M. Sze, "Physics of Semiconductor Devices," John Wiley & Sons, 1981.

27. D. Neamen, "Semiconductor Physics and Devices: Basic Principle, 3rd ed." New York: McGraw Hill, 2002.

28. K. K. Thornber, "Applications of scaling to problems in high-field electronic transport," *Journal Applied Physics*, vol. 52, no. 1, pp. 279-290, Jan. 1981.

29. K. G. McKay, "Avalanche breakdown in silicon," *Physical Review*, vol. 94, no. 4, pp. 877-884, May 1954.

30. R. A. Kokosa and R. L. Davies, "Avalanche Breakdown of Diffused Silicon p-n Junctions," *IEEE Transactions on Electron Devices*, vol. ED-13, no. 12, pp. 874-881, Dec. 1966.

31. C. G. B. Garrett and W. H. Brattain, "Some experiments on, and a theory of, surface breakdown," *Journal of Applied Physics*, vol. 27, no. 3, pp. 299-306, Mar. 1956.

32. O. Leistiko and W. W. Hooper, "Effect of Surface Fields on the Breakdown Voltage of Planar Silicon p-n Junctions," *IEEE Transactions on Electron Devices*, vol. ED-14, no. 3, pp. 157-162, Mar. 1967.

33. M. H. U. Habib and N. McFarlane, "Breakdown and optical response of CMOS perimeter gated single-photon avalanche diodes," *IEEE Electronics Letters*, vol. 53, no. 19, pp. 1323-1325, Sep. 2017.

A Novel One SWS-FET Transistor for AND/OR Logic Gate

Bander Saman[1,2], E. Heller[3,†] and F. C. Jain[1,*]

[1]*Department of Electrical and Computer Engineering, University of Connecticut,
371, Fairfield Way, U-4157, Storrs, CT 06269, USA*
[2]*Department of Electrical Engineering, Taif University,
P.O.BOX 888 - 21974- Hawiyah- Taif- KSA*
[3]*Synopsys Inc. 400 Executive Boulevard, Ossining, NY 10562, USA*
[*]*faquir.jain@uconn.edu*, [†]*evankheller@gmail.com*

This paper presents the design and modeling of AND/OR logic gate using one high-mobility n-channel spatial wave-function switched field-effect transistor (n-SWS-FET), which provide a significant reduction of cell area and power dissipation. In SWSFET, the channel between source and drain has two or more quantum well (QW) layers separated by a high band gap material between them. The gate voltage controls the charge carrier concentration in the two quantum well layers and it causes the switching of charge carriers from one channel to other channel of the SWS device. This switching property promises to build AND/OR logic gate with one n-SWS-FET transistor, where Complementary Metal Oxide Semiconductor (CMOS) AND/OR gate is built by 6 transistors. The proposed gate configures as AND/OR by change sources signal. The SWS-FET device with two well $Si/Si_{0.5}Ge_{0.5}$ has been modeled using Berkeley Short-channel IGFET Model (BSIM4.6.0) and Analog Behavioral Model (ABM), the model is suitable for transient analysis at circuit level. This model is optimized for AND/OR logic and used to replace a conventional CMOS logic.

Keywords: SWSFETs; quantum well; logic gates; VLSI.

1. Introduction

Unlike conventional FETs, the two quantum well n-SWS-FET is comprised of two vertically stacked coupled quantum well or quantum dot channels as shown in Fig. 1. Where W_1 is upper undoped Si quantum well and W_2 is lower undoped Si quantum well, the wells are sandwiched between two $Si_{0.5}Ge_{0.5}$ barriers [1, 2].

The quantum well current is transported between the source and drain through upper and lower Si quantum wells, where the current in upper well W_1 is I_{DS1} and the current in lower well W_2 is I_{DS2}. The upper and lower quantum well currents in n-SWS-FET are function of gate voltage (V_G). Once the V_G is increased above threshold voltage of lower quantum well W_2 (V_{TH2}), the electrons appear in lower quantum well, and as $V_G >$ threshold voltage of upper quantum well W_1 (V_{TH1}), the carriers start transferring to W_1. In a small gate voltage range, current flows in both quantum wells W_1 and W_2. Finally, as V_G is further increased, carriers completely transfer from the lower quantum well W_2 to the upper quantum well W_1, resulting in current flow only in W_1 [3, 4].

Fig. 2 shows the charge-densities simulation for the two $Si/Si_{0.5}Ge_{0.5}$ QWs n-SWS-FET. Besides, Fig. 3 shows the two $Si/Si_{0.5}Ge_{0.5}$ QWs n-SWS-FET energy band diagram of Type II heterostructure (strained $Si/Si_{0.5}Ge_{0.5}$), where the conduction band offsets $\Delta Ec = Ec_{Si} - Ec_{SiGe} = 0.15eV$ in strained $Si/Si_{0.5}Ge_{0.5}$ heterostructure. Fig. 4 presents quantum simulations showing the transfer of electron wave-function from W_2 to W_1 as the V_G is changed from 0.4 to 1.2 V.

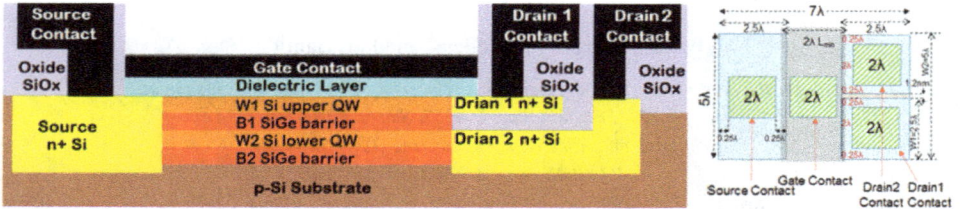

Fig. 1. Two $Si/Si_{0.5}Ge_{0.5}$ QWs n-SWS-FET Cross-sectional [2].

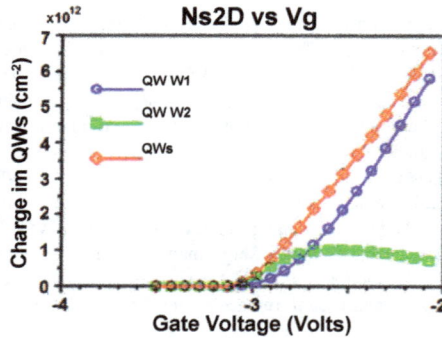

Fig. 2. Type II Heterostructure of two $Si/Si_{0.5}Ge_{0.5}$ QWs n-SWS-FET energy band diagram.

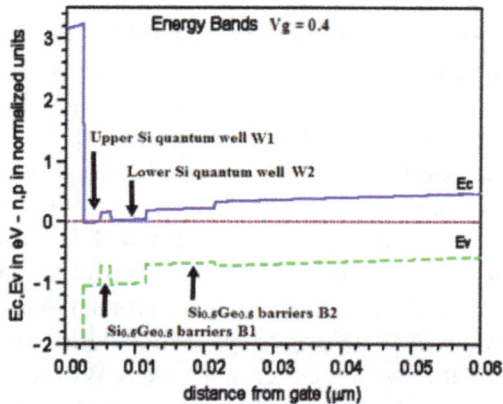

Fig. 3. Type II Heterostructure of two Si/Si0.5Ge0.5 QWs n-SWS-FET energy band diagram.

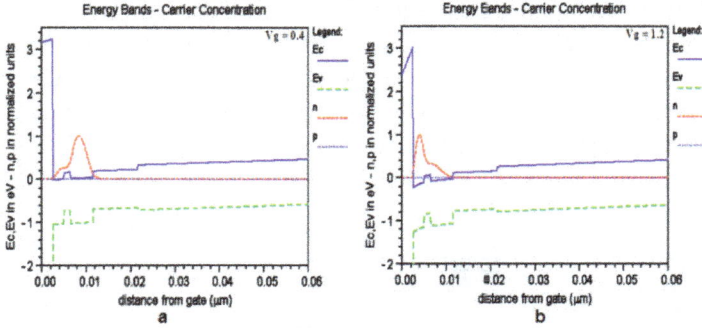

Fig. 4. Two Si/Si$_{0.5}$Ge$_{0.5}$ QWs n-SWS-FET electron wave-function in W2 at 0.4V (a) & W1 at 1.2V (b).

2. Two QWs n-SWS-FET Circuit Model

In terms of the device operation, the two-QW n-SWS-FET works as an enhancement n-channel metal-oxide field-effect transistor (NMOS-FET). Eq. (1) illustrates the drain current for an NMOS-FET, and this equation was modified for the two-QWs n-SWS-FET to represent the drain current for the upper quantum well W$_1$ and the drain current for lower quantum well W$_2$ in Eq. (2) and Eq. (3), respectively [2-4].

$$I_{DS} = \left(\frac{W}{L}\right) C_{OX}\mu_n \left((V_{GS} - V_{TH})V_{DS} - \frac{V_{DS}^2}{2} \right) \tag{1}$$

$$I_{DS1} = \left(\frac{W1}{L}\right) C_{OX}\mu_n \left((V_{GS1} - V_{TH1})V_{D1S1} - \frac{V_{D1S1}^2}{2} \right) \tag{2}$$

$$I_{DS2} = \left(\frac{W2}{L}\right) C_{OX}\mu_n \left((V_{GS2} - V_{DTH2})V_{D2S2} - \frac{V_{D2S2}^2}{2} \right) \tag{3}$$

$$V_{DTH2} = \begin{cases} V_{TH2} & \text{When } V_{GS} < V_{UL} \\ V_{TH2} + \frac{(V_{GS} - V_{UL})^2}{(V_{TH1} - V_{UL})} & \text{When } V_{GS} > V_{UL} \end{cases} \tag{4}$$

Where

L	the channel length
W1	the upper QW width
W2	the lower QW width
C_{OX}	the capacitance per unit area
μ_n	the electron mobility
V_{GS1}	the voltage between gate to source of the upper QW
V_{GS2}	the voltage between gate to source of the lower QW
V_{D1S1}	the voltage between drain to source of the upper QW
V_{D2S2}	the voltage between drain to source of the upper QW
V_{DTH2}	the developed threshold voltage of the lower QW
V_{TH1}	threshold voltage of upper the upper QW
V_{TH2}	the threshold voltage of lower QW
V_{UL}	the transition voltage

BSIM4.6 and ABM libraries are integrated to establish the two-QW n-SWS-FET model that can address the switching and changing in threshold voltages [4-6]. Fig. 5 shows the two-QW n-SWS-FET circuit model. The model is used in Cadence-OrCAD CIS simulator as hierarchical circuit. Fig. 6 shows I_{DS}-V_{GS} characteristics of the two-QW n-SWS-FET, where Table 1 presents the circuit model configuration.

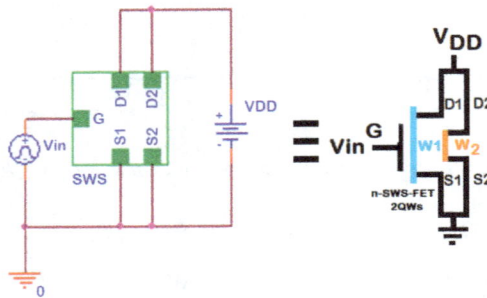

Fig. 5. The Cadence hierarchical circuit model for two QWs n-SWS-FET.

Table 1. Two-QW n-SWS-FET Parameters.

Parameter	Value	Unit
Channel length L	20	nm
Upper well W1 width	40	nm
Lower well W2 width	100	nm
V_{DD}	1	V
V_{TH1}	0.3	V
V_{TH2}	0.2	V
V_{UL}	0.025	V

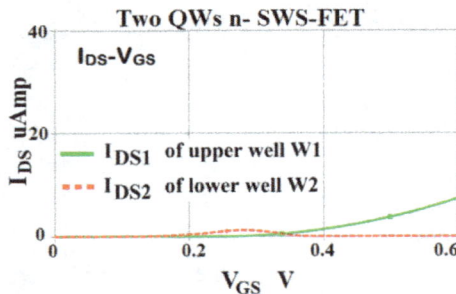

Fig. 6. The simulated I_{DS}-V_{GS} of 20 nm two QWs n-SWS-FET.

3. Operation Mode of Two QWs n-SWS-FET

NMOS-FET make very good two state electronic switches in CMOS digital circuits as they operate between open (OFF) and closed (ON) switching modes. The open occurs at cutoff region (non-conducting) and the closed happens at linear or saturation region (conducting or low resistance between drain and source) [7].

Based on the carriers transferring, the two-QW n-SWS-FET device has four switching modes as shown in Table 2 and Fig. 7. When $V_G > V_{TH2}$, both QWs W_1 and W_2 are in off mode ($I_{DS1} \approx 0$ and $I_{DS2} \approx 0$). As $V_G > V_{TH2}$, W_1 is still in off mode and W_2 is in off mode ($I_{DS1} \approx 0$ and $I_{DS2} > 0$). Once $V_G > V_{TH1}$, the electrons transfer from W_2 to W_1, and the current flows in W1 ($I_{DS1} > 0$) and I_{DS2} drop off. And at time of $V_G \gg V_{TH1}$, W_1 is in on mode and W_2 is in off mode ($I_{DS1} \gg 0$ and $I_{DS2} \approx 0$) [4].

Table 2. The operation mode.

V_G	$V_G < V_{TH2}$	$V_G > V_{TH2}$	$V_G > V_{TH1}$	$V_G \gg V_{TH1}$
Well 1	Off mode, $I_{DS1} \approx 0$	Off mode, $I_{DS1} \approx 0$	On mode, $I_{DS1} > 0$	On mode, $I_{DS1} \gg 0$
Well 2	Off mode, $I_{DS2} \approx 0$	On mode , $I_{DS2} > 0$	$\approx$ Off mode, $I_{DS2} \to 0$	Off mode, $I_{DS2} \approx 0$

In two-QW n-SWS-FET's, the upper well (W_1) behaves like NMOS-FET; W_1 is as open switch (non-conducting) when $V_G < V_{TH1}$ and closed (conducting D_1 with S_1) when $V_G > V_{TH1}$. The lower well (W_2) is as an open switch (non-conducting) at $V_G < V_{TH2}$ or $V_G \gg V_{TH1}$, and closed (conducting D_2 with S_2) at $V_{TH1} \gg V_G > V_{TH2}$.

By counting W_1 and W_2 switching modes together, a two-QW n-SWS-FET can be used as a four-state switch 00-01-11-10 as shown in Fig. 7. Furthermore, the two-QW n-SWS-FET is suitable for use with the 4 position selector switch as shown in Table 3, where the selected well is marked as X in the table.

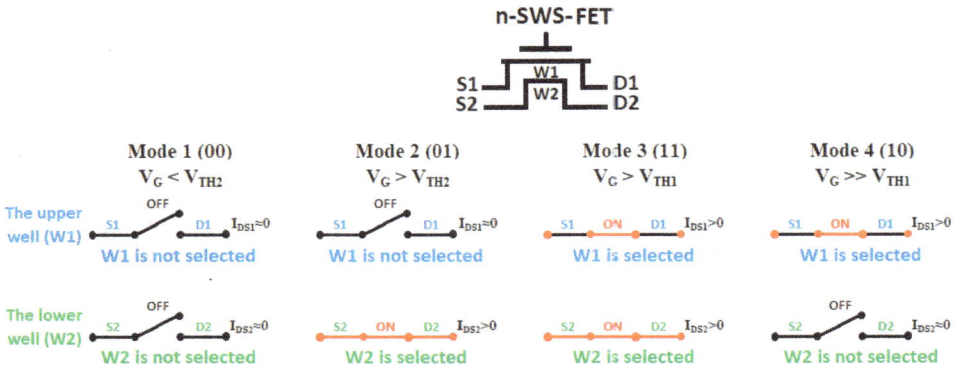

Fig. 7. The simple circuit for switching operation modes in two QWs n-SWS-FET.

Table 3. The four position selector switch in two QWs n-SWS-FET

Gate Voltage V_G	The lower well (W_2)	The upper well (W_1)
$V_G < V_{TH2}$		
$V_G > V_{TH2}$	X	
$V_G > V_{TH1}$	X	X
$V_G \gg V_{TH1}$		X

4. One n-SWS-FET Transistor as AND/OR Logic Gate

The input/output of MOSFETs is high (V_{DD}) or low (0 = GND). PMOS-FET will only be ON if the V_G is low and the source voltage (V_S) is connected to VDD. Likewise, NMOS-FET will only be ON when V_G is high and V_S is connected to GND. In CMOS networks, the output is 'pull-up' to V_{DD} if there is a connected path in PMOS-FET network, On the other hand, the output is 'pull-down' to 0 if there is a connected path in NMOS-FET network. Fig. 8 shows the Pull-up and Pull-down networks in circuit I.

In circuit II-Fig. 8, the value of NAND gate (T1, T2, T3, T4) is low only if both A and B are high, because the Pull-down network has the path to ground only when T1 and T2 are ON. As the output of NAND passes through the inverter logic (T5, T6), the output state of the AND logic gate only high when of its inputs are at a high level (V_{DD}).

Similarly, the circuit III-Fig. 8 shows the circuit of OR gate. When A and B are low the Pull-up network has the path to V_{DD}, which put the NOR gate (T7, T8, T9, T10) in high state and the OR logic gate in low level (0).

The design of two inputs (A,B) AND gate using two-QW n-SWS-FET is less complicated as shown in Fig. 9, only one transistor is used in SWS-AND gate. The input A is connected to the gate and the input B is connected to the upper well W_1 source, where the lower well W_2 source connects to the ground. The proposed one transistor SWS-AND gate has no power supply, thus it can be referred as the Powerless gate.

To reiterate the above description of two-QW n-SWS-FET switching characteristics, if the input signal is low ($V_A < V_{TH1}$), then the output will be low because either no wells selected or W_2 is selected as shown in Fig. 9 case 1 and 2. When V_A signal is high ($V_A > V_{TH1}$), The W_1 is selected which allows V_B signal to pass to the output as shown in Fig. 7 case 3 and 4 [at V_B=1 the output is high, at V_B=0 the output is low].

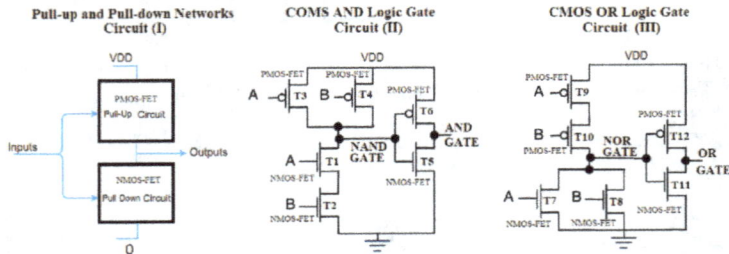

Fig. 8. The Pull-up and Pull-down Networks (I), COMS AND Logic Gate (II), CMOS OR Logic Gate (III).

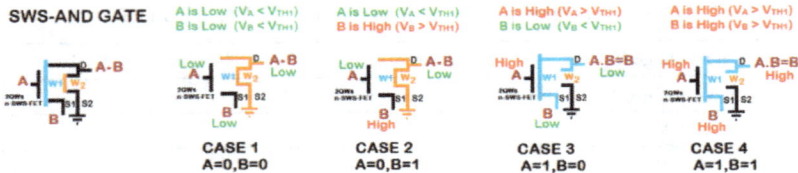

Fig. 9. The SWS-AND Logic Gate.

SWS-AND gate can be configured to perform SWS-OR gate by connecting the source as shown in Fig. 10. Where, the upper well source connects to V_{DD} and the lower source connects to B signal. The proposed one transistor OR gate has no ground and can be referred as Groundless gate.

When V_A is high, the upper well is selected and the output corresponds to the upper source signal which is V_{DD}, this is presented in Fig. 10 case 1 and 2. If V_A is low the upper well is selected and the output = V_B as shown in Fig. 10 case 3 and 4.

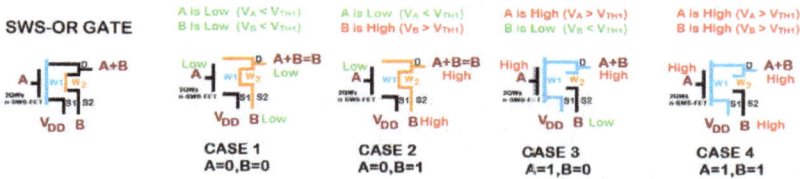

Fig. 10. The SWS-OR Logic Gate.

Fig. 11 shows SWS-AND/OR gate layouts (twin source configuration). The minimum n-SWS-FET area is $5\lambda \ast 7\lambda$ (Width*Length), where λ is a scale factor and the minimum technology geometry. It follows that the area of SWS-AND/OR gate is clearly less than the area of CMOS AND/OR gate as shown in Fig. 12. The minimum CMOS AND/OR gate area is $12\lambda \ast 26\lambda$.

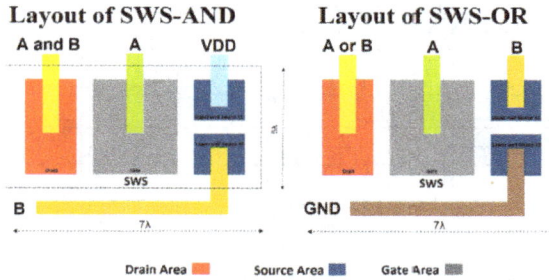

Fig. 11. The SWS-AND/OR Gate Layout.

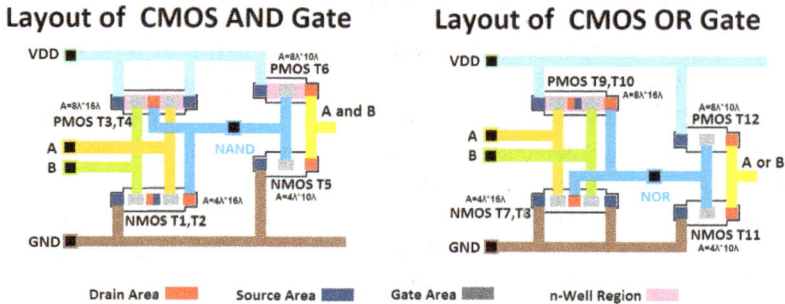

Fig. 12. The CMOS-AND/OR Gate Layout.

5. SWS-AND/OR Logic Gates Simulation

The accuracy of the SWS-AND/OR logics is verified by SWS-FET model in Cadence. The channel length of n-SWS-FET is 20nm; the low inputs state is $\approx 0V$ while the high inputs state is $\approx 1V$ which is V_{DD}. The simulation of 20 nm SWS-AND gate and 20 nm SWS-OR gate are shown in Fig. 13.

  Here, the inputs V_A and V_B are two square pulse signal providing all four possibilities. As expected, the output of the SWS-AND gate is high ($\approx 1V$) when V_A=V_B=1V, otherwise the output is low (≈ 0), which is indicating a correct AND logic. For the SWS-OR gate, the output is low (≈ 0) when V_A=V_B=0V, otherwise the output is high ($\approx 1V$). Moreover, the outputs (dash line) of 20 nm SWS-AND gate and 20 nm SWS-OR gate are shown as dash line in Fig. 11. The outputs of the CMOS AND/OR and the SWS-AND/OR gates are identical and it is obvious that SWS-AND/OR can be used as a novel one SWS-FET transistor for AND/OR logic gate.

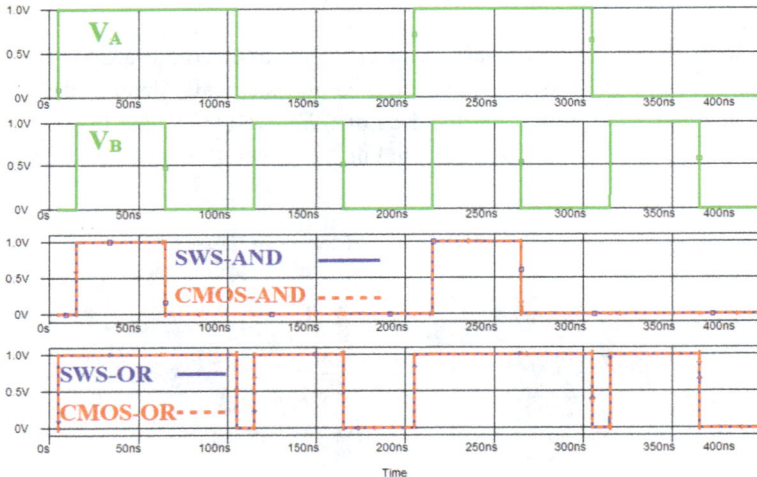

Fig. 13. The simulation of CMOS-AND/OR gate and SWS-AND/OR gate.

6. Average Power Consumption Analysis of SWS-AND/OR Logic Gates

Power consumption is the subject for recent VLSI circuits [8]. In this work the power consumption of SWS-AND/OR 20nm technology gate is measured and compared with conventional CMOS-AND/OR. Fig. 14 shows the transient simulation of one cycle period of operation (thus having four possible input A, B signal 11-01-10-00) and the power consumptions for CMOS-AND-gate and SWS-AND-gate. Fig. 15 illustrates the simulation result of CMOS-OR-gate and SWS-OR-gate. The average of one cycle (40ns) power consumption across CMOS-AND, SWS-AND, CMOS-OR and SWS-OR are 0.906 µWatts, 0.0316 µWatts, 0.434 µWatts and 0.012 µWatts, respectively. Fig. 16 shows the graphical representation of the average power consumption. Table 3 is a comparison of the area for CMOS-AND and SWS-AND.

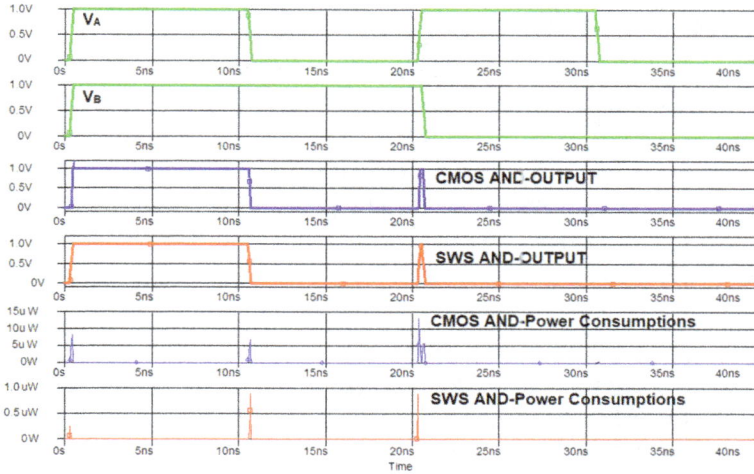

Fig. 14. The power consumptions simulation of CMOS-AND gate and SWS-AND gate.

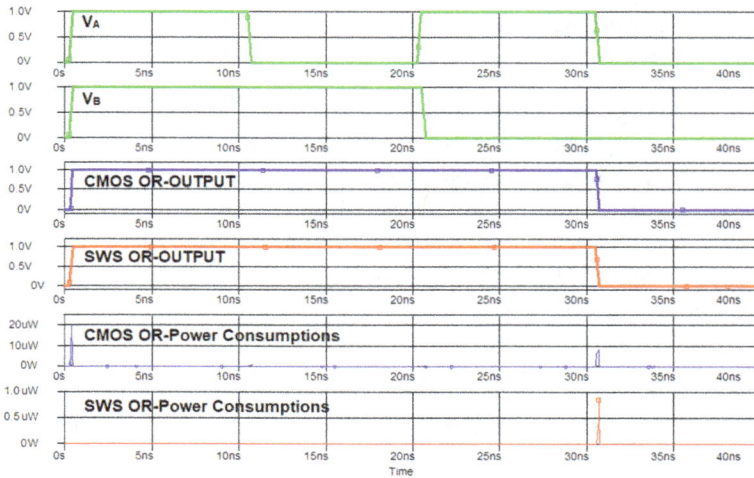

Fig. 15. The power consumptions simulation of CMOS-OR gate and SWS-OR gate.

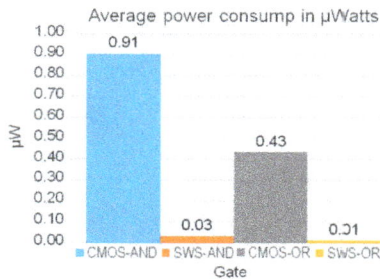

Fig. 16. The Average power consumption of CMOS-AND/OR and SWS-AND/OR gates.

7. Conclusions

Using one n-SWS-FET transistor (two-QW n-SWS-FET) for AND/OR gate reduces the number of transistors by 1-1/6 = 83% compared with CMOS gate. SWS-AND/OR circuits have the same accuracy as that of CMOS AND/OR circuits. The proposed one transistor SWS-AND/OR gate have only either VDD or GND, thus preventing the flow of short circuit current. This is a novel small area and flexible AND/OR circuit. SWS-AND/OR gate has almost 9 times less area. The simulation results show that the power consumed by the SWS-AND/OR circuit is only about 70% when compared to CMOS-AND/OR circuits.

References

[1] F. Jain, J. Chandy, and E. Heller, Int. J. High Speed Electronics and Systems, Vol. 20, 9 (2011).
[2] P. Gogna, M. Lingalugari, J. Chandy, E. Heller, E-S. Hasaneen and F. Jain, "Quaternary Logic and Applications Using Multiple Quantum Well Based SWSFETs", Int. Journal of VLSI Design & Communication Systems, Vol. 3, No. 5, 11 (2012).
[3] B. Saman, P. Mirdha, M. Lingalugari, P. Gogna, F. C. Jain, El-Sayed Hasaneen, and E. Heller, "Logic Gates Design and Simulation Using Spatial Wavefunction Switched (SWS) FETs", International Journal of High Speed Electronics and Systems, 9 (2015), Vol. 24, No. 03n04.
[4] B. Saman, P. Gogna, E-S. Hasaneen, E. Heller and F.C. Jain, "Spatial Wavefunction Switched (SWS) FETs SRAM Circuits and Simulation", International Journal for High Speed Electronics and Systems (IJHSES), Vol. 26, No. 3 (2017).
[5] Neeraja Jagadeesan, B. Saman, M. Lingalugari, P. Gogna, and F. Jain, "Sequential Logic Circuits Using Spatial Wavefunction Switched (SWS) FETs", International Journal of High Speed Electronics and Systems, 9 (2015), Vol. 24, No. 03n04.
[6] Saman, B., Kondo, J., Chandy, J., and Jain, F. C., "Circuits and Simulation of Quaternary SRAM Using Quantum Dot Channel Field Effect Transistors (QDC-FETs)", International Journal of High Speed Electronics and Systems, 27(01n02).
[7] Pashmineh, S., and Killat, D. "High-voltage circuits for power management on 65 nm CMOS", Advances in Radio Science, 13, 109-120.
[8] Ahmed Shariful Alam, Md. Abdur Rahman, Abu Hena Md. Mustafa Kamal, and Md Salek Mahmud, "SET Based Low Power Consuming Digital Circuits", 8th International Conference on Electrical and Computer Engineering, 20-22 December 2014, Dhaka, Bangladesh.

Multi-State 2-Bit CMOS Logic Using n- and p-Quantum Well Channel Spatial Wavefunction Switched (SWS) FETs

Faquir Jain[1,*], Bander Saman[1,2], Raja Hari Gudlavalleti[1], John Chandy[1] and Evan Heller[3]

[1]*Department of Electrical and Computer Engineering, University of Connecticut, 371 Fairfield Way, Unit 4157, Storrs, CT 06269, USA*
[2]*Department of Electrical Engineering, College of Engineering, Taif University, Saudi Arabia*
[3]*Synopsys Inc., Ossining, NY 10562 USA*
*faquir.jain@uconn.edu

Unlike conventional FETs, spatial wavefunction switched (SWS)-FETs are comprised of two or more vertically stacked coupled quantum well or quantum dot channels, and the spatial location of carriers within these channels is used to encode the logic states (00), (01), (10) and (11). The aim of this paper is to present 4-states/2-bit output-input transfer characteristics using two Si/SiGe quantum well channels configured as CMOS using n- and p-channel spatial wavefunction switched field-effect transistors (SWS-FETs). Quantum simulations show switching of wavefunctions as the gate voltage is increased from lower Si quantum well to the upper well in n-channel and from upper SiGe quantum well to lower well in the p-channel. The inverter transfer characteristic and current switching are obtained by integrating BSIM (Berkeley Short-channel IGFET Model) and the Analog Behavioral Model (ABM). The simulation shows current flow only during switching.

Keywords: Spatial wavefunction switched FET; SWS-FETs; CMOS-SWS; quantum well/dot channel; quantum nanosheets.

1. Introduction

The spatial wave-function switched field-effect transistor (SWS-FET) has two or more vertically stacked quantum well/dot channels. The magnitude of gate voltage determines the location of electrons in lower or upper quantum well channels. The spatial location of carriers in multiple quantum well channels is used to encode the logic states (00), (01), (10) and (11) [1]. Since each channel can have its own source and drain, the current in a SWS-FET is routed between multiple channel drains according to the applied voltage on a single gate. In term of circuits, the inherent multi-state operation of SWS-FETs provides opportunity to process 2-bit simultaneously. We have reported logic cells, full-adders, latches and static random access memory (SRAM) cells using 4-state/2-bit SWS-FETs with an area saving of 50% [1-5]. These circuits have utilized multiple source/drain configurations of n-channel SWS-FETs, enabling processing of 2- or more bit simultaneously. This paper presents the use of both n- and p-channel SWS-FETs to implement 4-state/2-bit complementary metal oxide semiconductor (CMOS) logic, which reduces the power dissipation while preserving the reduction of FET count. SWS channels may be formed either with quantum wells or with coupled quantum dot superlattices (QDSL).

2. Twin-Drain n-Channel SiOx-Si Cladded Quantum Dot SWS-FET: Experimental

Figure 1(a) shows the schematic cross-section of a fabricated n-SWS-FET structure and Figure 1(b) illustrates its ID-VD characteristics [1, 2]. The device has four layers of SiOx-cladded Si quantum dots serving as the transport channel. It has two drains (D2-deep, D1-shallow). The two lower QD layers are connected to drain D2 and upper two QD layers are interfaced to shallow drain D1 [2]. Figure 1(b) shows the voltage-current characteristics at gate voltages (VG) of 2.25 and 2.5 V, respectively. The lower dot layers are conducting for a drain voltage in the range of 1.8-2.0 V whereas the upper channel is conducting from 2.2-2.4 V [1].

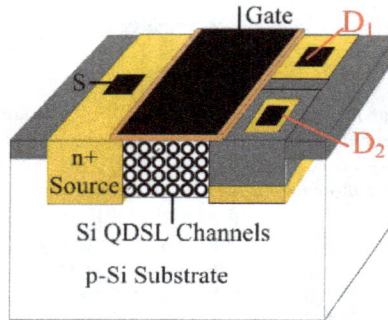

Fig. 1(a). Four quantum dot layers forming two channels of a twin-drain n-SWS-Si FET.

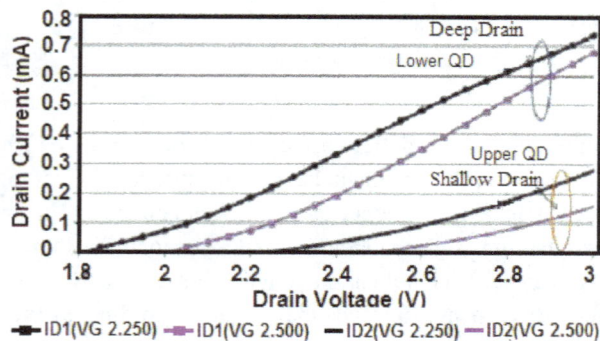

Fig. 1(b). Experimental ID-VD characteristics showing conduction in lower and upper channels.

Wavefunction switching has also been experimentally observed in 2-well and 4- well InGaAs SWS devices as shown in capacitance-voltage plots of Figure 2. Gate capacitance depends on the location of carriers. Hence, multiple C-V peaks indicate transfer of carriers from lower wells to the upper wells. An asymmetrically coupled quantum well InGaAs

SWS-MOS device shown in Figure 1(c), fabricated using II-VI gate insulator, has exhibited the distinct capacitance peak around gate voltage of 0 V. This corresponds to charge in the lower well W2 as we move away from inversion towards accumulation. The transfer to upper well W1 takes place at a voltage of -2.0 V. Simulation results have also shown two peaks, one near the accumulation (left ~ -4.0 V) and the other near the inversion regime (right ~ -2.8 V) [1].

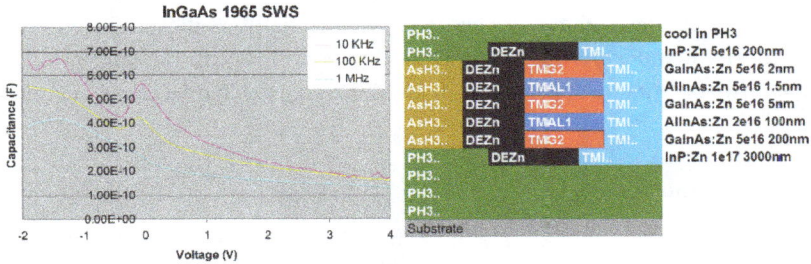

Fig. 2. Capacitance vs Vg in two InGaAs quantum well and AlInAs barrier SWS-MOS.

3. CMOS SWS-FET Quantum Well Structure

The complementary SWS-FET is shown in Fig. 3 where the Si quantum wells and SiGe barriers layers are epitaxially grown on p-Si and n-Si regions sharing a common substrate. Here, two Si quantum wells (W1 upper well, W2 lower well) are sandwiched between SiGe barriers. The two devices are separated by an oxide/insulator layer.

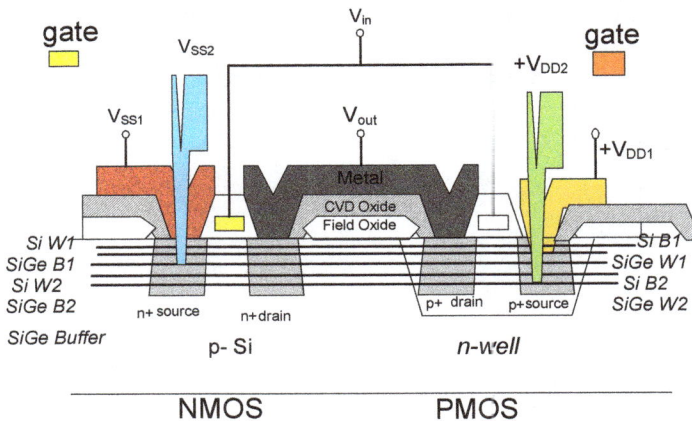

Fig. 3. CMOS device having an n-channel and a p-channel SWS-FET.

The Si and SiGe layers serving as quantum wells and barriers are grown on relaxed $Si0.75Ge0.25$ buffer layer. $Si0.5Ge0.5$ layer is tensile strained and forms a Type II heterostructure with Si layer. The n-FET is grown on p-Si tub/well and p-FET is realized on n-Si tub/well.

In the case of n-channel SWS FET, the upper Si quantum well W1 is adjacent to gate insulator (e.g., ZnMgS, HfO2) on one side and SiGe barrier on the other. However, for p-channel SWS FET, Si layers serves as barriers for the upper well W1 which is Si0.5Ge0.5. Once the input voltage or gate voltage Vg is increased above threshold voltage VTH2, the electrons appear in lower quantum well W2, and as the gate voltage is increased further, the carrier (or their wavefunction) start transferring to the upper well W1 for n-channel SWS FET. This is due to well W2 having a larger thickness. In a small voltage range, current flows in both lower and upper quantum wells. Finally, as the gate voltage is further increased, carrier completely transfer from the lower well W2 to the upper well W1 and the current flows only in the upper well. This is illustrated using quantum simulation of the structure of Fig. 3.

4. Quantum Simulation

Figure 4(a) shows electron wavefunction in an n-channel SWS FET in lower Si quantum well W2 at 0.2V when the input voltage (Vin=Vgs) exceeds threshold voltage VTH2. The wavefunction transfers to the upper Si quantum well W1, shown in Fig. 4(b) as the gate voltage is increased from 0.2 to 0.8 V (VTH1).

Unlike electron wavefunctions which are located in Si wells, hole wavefunctions are located in SiGe layers which serve as the quantum wells. Note that the Si-SiGe system forms a type II heterostructure where QWs are tensile strained. Figure 5(a) shows hole wavefunctions in lower SiGe well W2 for gate voltage of -0.4 V. Holes transfer to the upper SiGe well W1 at a gate voltage Vg = -0.8V. Here, the gate voltage is with respect to source (VDD) of p-channel FET. An input voltage (Vin or Vg) of 0.2 V with respect to ground translates to Vg = -.8V for p-channel FET with respect to its source or VDD.

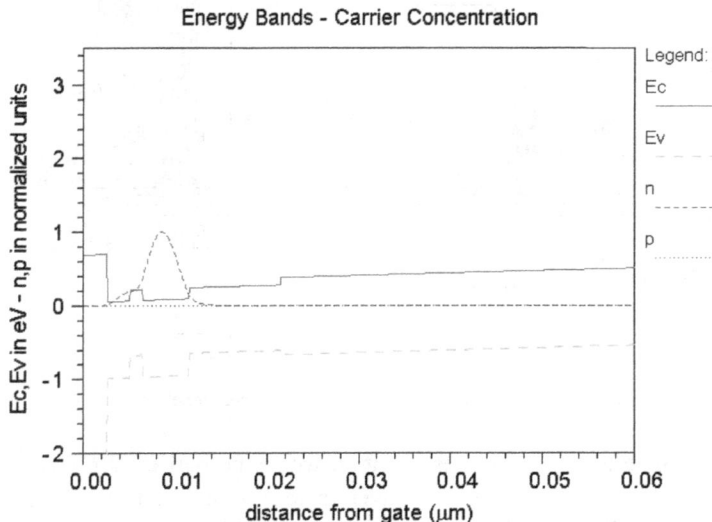

Fig. 4(a). Electron wave-function in lower quantum well W2 of n-channel FET at Vg=0.2V(a).

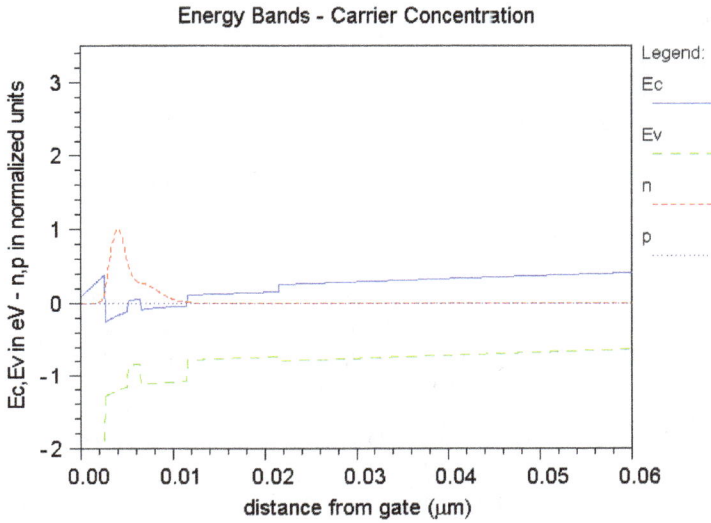

Fig. 4(b). Electron wavefunction in n-channel switched to upper quantum well W1 at Vg=0.8V.

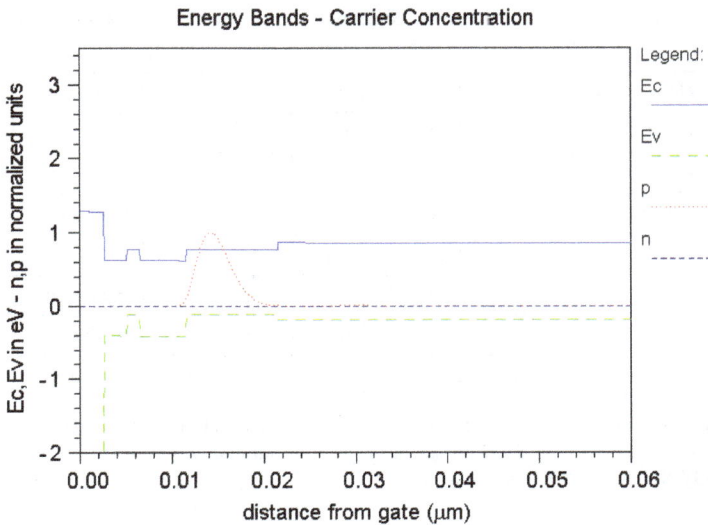

Fig. 5(a). Hole wavefunction at Vg = -0.4V switched to lower well W2 in a p-channel SWS-FET.

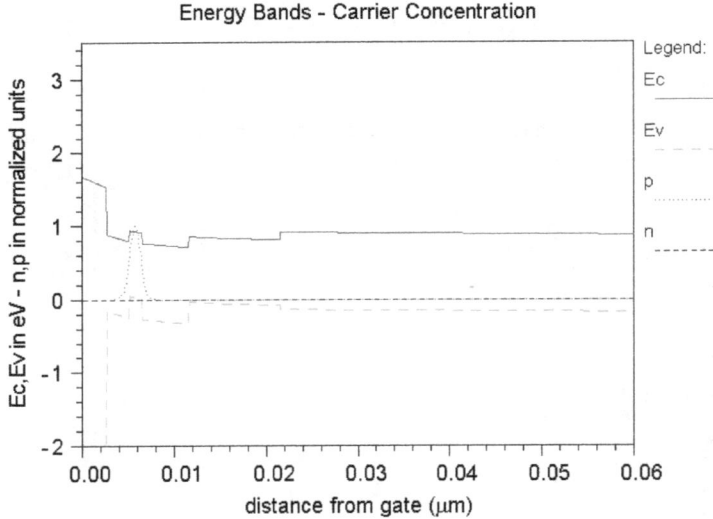

Fig. 5(b). Hole wavefunction confined in upper well W1 at Vg = -0.8V.

Table I shows the values of the parameters used for quantum simulations of Figures 4 and 5. The effect of electron affinity χ change has been observed and may be used to adjust relative values of voltage thresholds.

Table I. Parameters used in the simulation of wavefunction for two QW channel-SWS-FET [Si-SiGe-Type II-SWS Structure: (Vg=0.8, 0.2), p-channel (-0.4,-0.8)].

Layer	Thick (um)	χ (eV)	E_g (eV)	m_e	m_h	εr	N_d (cm^{-3})	N_a (cm^{-3})
ZnS	0.0025	3.5	3.80	0.13	0.38	8.0	0.00E+00	0.00E+00
Si (QW1)	0.0025	4.15	1.04	0.19	0.49	11.9	0.00E+00	0.00E+00
SiGe(.5)	0.0015	4.0	0.89	0.13	0.38	14	0.00E+00	0.00E+00
Si (QW2)	0.0050	4.15	1.04	0.19	0.49	11.9	0.00E+00	0.00E+00
SiGe(.5)	0.0100	4.0	0.89	0.13	0.38	14	0.00E+00	0.00E+00
SiGe(.75)	0.0500	**3.9**	1.05	0.13	0.38	14	0.00E+00	1.00E+16
Si	0.1000	3.8	1.1	0.19	0.49	11.9	0.00E+00	1.00E+16

* QWs are tensile strained, X is electron affinity, E_g the band gap, me and m_h are the electron and hole masses, ε_r the dielectric constant, N_d and N_a are donor and acceptor concentrations). V_{gs} for n-SWS-FET are 0.4 V and 1.2 V, V_{gs} for p-SWS-FET are -0.2 V and -0.8 V.

The hole mobility in SiGe quantum wells is higher by a factor of 2 as compared to Si quantum wells (e.g. hole mobility for $Si_{0.5}Ge0.5$ is 1480 cm^2/V-s). This results in aspect ratio (W/L) of both n-SWS-FET and p-SWS-FET to be similar.

5. SWS-FET Circuit Model and Logic

In the two-QW n-channel SWS-FET, when the gate to source voltage (V_{GS}) is applied between 0 and below threshold of lower well (V_{th2}), both wells W1 and W2 are in an off mode so both currents in lower well (I_{DS2}) and upper well (I_{DS1}) are zero. As the gate voltage is set above V_{th2}, the electrons are confined in well W2, resulting in drain current, IDS2, flowing in W2. As V_{GS} becomes greater than the threshold voltage of the upper well (V_{th1}), the electrons transfer from W2 to W1, drain current IDS1 flows in W1, and I_{DS2} drops off. At first transition voltage (V_{UL}), W2 of n-channel SWS is in an off mode and W1 of p-channel SWS is in ON mode [1-4]. The output voltage is expressed as following five states:

$$V_{out} = V_{DD1}$$

$$V_{out} = \frac{(V_{DD1} - V_{SS2})(R_{n2})}{R_{p2} + R_{n2}}$$

$$V_{out} = \frac{(V_{DD2} - V_{SS2})(R_{n2})}{R_{p2} + R_{n2}} = \frac{V_{DD2} - V_{SS2}}{2}$$

$$V_{out} = \frac{(V_{DD2} - V_{SS1})(R_{n1})}{R_{p2} + R_{n1}} \approx (V_{DD2} - V_{SS1}) * \frac{R_{n1}}{R_{p2}}$$

$$V_{out} = V_{SS1}$$

This can be understood using Fig. 6(a) schematically showing SWS-quantum well channels and Fig. 6(b) showing its equivalent circuit. The value of resistance is smaller if the subscript is '1' for upper well 1.

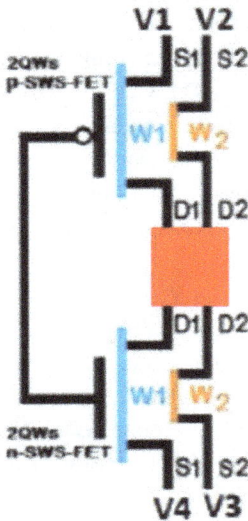

Fig. 6(a). CMOS configuration SWS-FETs.   Fig. 6(b). Equivalent circuit showing resistors for various channels.

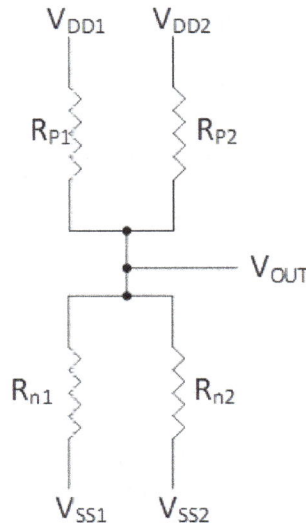

Tables II and III summarizes the operation modes of n-SWS-FET and p-SWS-FET, respectively.

Table II. The switching mode for two QW n-SWS-FET.

Gate voltage V_G	$V_{GS} < V_{th2}$	$V_{GS} > V_{th2}$ $V_{GS} < V_{UL}$ $V_{GS} < V_{th1}$	$V_{GS} > V_{th2}$ $V_{GS} > V_{UL}$ $V_{GS} > V_{th1}$	$V_{GS} \gg V_{th2}$ $V_{GS} \gg V_{UL}$ $V_{GS} \gg V_{th1}$
Well-1	Off mode $I_{DS1} \approx 0$	Off mode $I_{DS1} \approx 0$	On mode $I_{DS1} > 0$	On mode $I_{DS1} \gg 0$
Well-2	Off mode $I_{DS2} \approx 0$	On mode $I_{DS2} \gg 0$	$\approx$ Off mode $I_{DS2} \to 0$	Off mode $I_{DS2} \approx 0$

Table III. The switching mode for two QW p-SWS-FET.

Gate voltage V_G	$V_{GS} \ll V_{th1}$ $V_{GS} \ll V_{UL}$ $V_{GS} \ll V_{th2}$	$V_{GS} < V_{th1}$ $V_{GS} < V_{UL}$ $V_{GS} < V_{th2}$	$V_{GS} > V_{th1}$ $V_{GS} > V_{UL}$ $V_{GS} < V_{th2}$	$V_{GS} > V_{th2}$
Well-1	On mode $I_{SD1} \gg 0$	On mode $I_{SD1} > 0$	Off mode $I_{SD1} \approx 0$	Off mode $I_{SD1} \approx 0$
Well-2	Off mode $I_{SD2} \approx 0$	$\approx$ Off mode $I_{SD2} \to 0$	On mode $I_{SD2} \gg 0$	Off mode $I_{SD2} \approx 0$

SWS-FET model is based on the integration of Berkeley Short-channel IGFET Model (BSIM4.6) and the Analog Behavioral Model (ABM). The two quantum well n-channel SWS-FET is modeled by combining two BSIM equivalent circuits (BSIM-EC) as shown in Figure 6(a) [5-6]. ABM blocks are used to represent threshold for lower quantum well W2 and upper quantum well W1 as well as all capacitances and current sources [7]. The resistances (shown in Fig. 6(b)) are calculated based on device geometry [8]. Drain current sources IDS2 (lower well W2) and IDS1 (upper well W1) are obtained from BSIM transistor model [1-3, 6]. The capacitances (e.g., CGB, CGD1, CGS1, CGD2, CGS2) are obtained from Meyer's capacitance model [9-10]. Cadence simulations are done using Enz-Krummenacher-Vittoz (EKV v301.01) model in 180 nm technology [16-20].

The CMOS-SWS inverter is modeled by two EKV transistors T1 and T2, where T1 and T2 represent upper well (W1) and lower well (W2) respectively. These have two threshold voltages VTH1 and VTH2 for T1 and T2 respectively.

The parameters of n channel SWS-FETs are (VTH1=0.4 V, VTH2=0.2 V, L=180 nm, W1=225 nm, and W2=450nm). The parameters of p channel SWS-FETs are (VTH1=-0.4 V, VTH2=-0.2 V, L=180 nm, W1=0.9 um, and W2=1.8 um). According to the EKV 180 nm model, the minimum W/L ratio of pSWS to nSWS is Kp/Kn=390/82=4.1. In order to program the inverter for 4-state operation, the circuit has to be connected to different VDD and VSS. These are: VDD1 (e.g, 1.8 V) value of upper W1 channel of p-SWS; VDD2 (=VDD1*2/3=1.2V); VSS2 (= VDD2*1/3=0.6V) for lower quantum well W2 of n-SWS; and VSS1 (=GND=0) for upper W1 of n-SWS. Fig. 7 shows BSIM-EC simulation of transfer characteristics of CMOS SWS-FET inverter using two quantum well n- and p-channels. Four-states are evident for the first time.

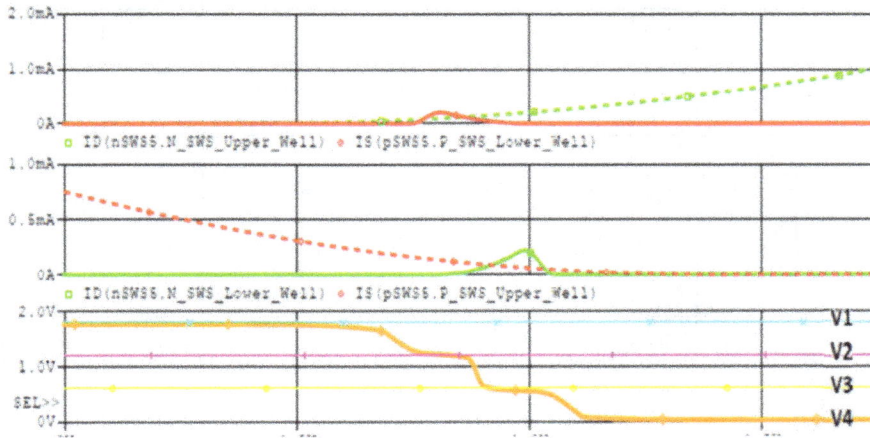

Fig. 7. BSIM-EC twin drains 20nm n- and p- SWS-FET modeled I_{DS-W2} (--) and I_{DS-W1} (—).

6. Comparison with Conventional CMOS

Table IV shows the comparison between quaternary CMOS-SWS and conventional CMOS [21, 22] inverter shown in Fig. 8. The simulation is shown in Fig. 9.

Table IV. Comparison of SWS-based (L=180nm, W=300nm) and conventional CMOS quaternary inverters.

Parameters	SWS	CMOS
The number of devices	2	6
Dynamic power 40nsec at logic 0,0.4V,0.8V,1.2V	0.54uW	8.2uW
Delay at logic change from 0 to 0.4V	10ps	34ps
Delay at logic change from 0.4V to 0.8V	4ps	10ps
Delay at logic change from 0.8V to 1.2V	12ps	26ps

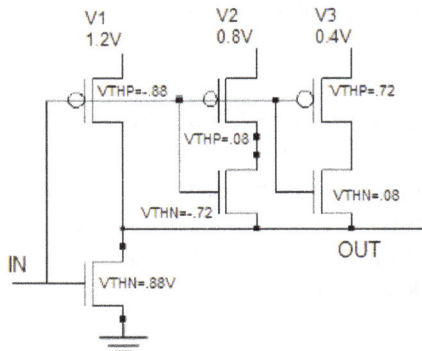

Fig. 8. CMOS Quaternary inverter circuit.

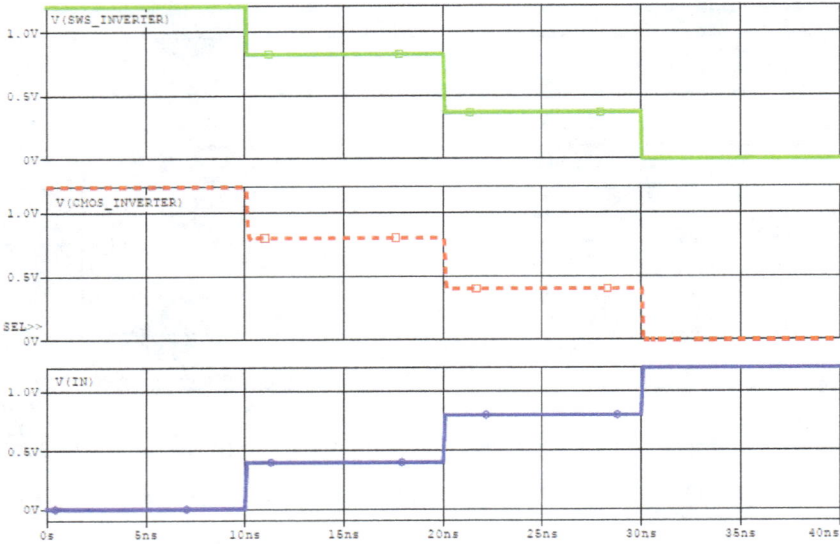

Fig. 9. Input (blue) and output waveform of 180nm SWS (green) and conventional (red) CMOS quaternary inverter.

7. Conclusion

This paper presents simulations of 4-state/2-bit CMOS SWS inverter using Si-SiGe quantum well channels. Preliminary experimental results for twin-drain SWS-Si FET based on QDSL has also been presented. The SWS transistor models are created by two BSIM equivalent circuits and simulated using Cadence software for 180 nm technology. The simulation shows that CMOS-SWS two channel quantum well devices has four states and low power resulting in increased performance and die area as compared to conventional CMOS quaternary inverter. The appropriate selection of the aspect ratio of upper and lower n- and p- channels, we believe to obtain 4-state using only VDD1 and GND. QDSL channels present an additional pathway to realizing stacked quantum dot based nanosheets. Nanosheets are being investigated in wrap-around FET structures [23].

Acknowledgments

This work was in part supported by Office of Naval Research contracts (N00014-02-1-0883 and N00014-06-1-0016), National Science Foundation grants (NER and ECS 0622068), and Connecticut Innovation grant. The authors gratefully acknowledge the assistance of Dr. Barry Miller, Prof. T-P. Ma, Mr. C-C. Yeh, L. Song and Mr. C. Tillinghast of Yale University for permitting the use of their laboratory facilities. The authors would like to further acknowledge Murali Lingalugari and Pial Mirdha of University of Connecticut.

References

1. F. Jain, M. Lingalugari, B. Saman, P.-Y. Chan, P. Gogna, E.-S. Hasaneen1, J. Chandy, and E. Heller, "Multi-State Sub-9 nm QDC-SWS FETs for Compact Memory Circuits," 46th IEEE Semiconductor Interface Specialists Conference (SISC), Atlanta (VA), December 2-5, 2015.

2. F. Jain, M. Lingalugari, J. Kondo, P. Mirdha, E. Suarez, J. Chandy, and E. Heller, "Quantum Dot Channel (QDC) FETs with Wraparound II–VI Gate Insulators: Numerical Simulations," Journal of Electronic Materials, vol. 45, no. 11, pp. 5663-5670, Mar. 2016.

3. S. Karmakar, John A. Chandy, and Faquir C. Jain. "Unipolar Logic Gates Based on Spatial Wave-Function Switched FETs." IEEE Transactions on Very Large Scale Integration (VLSI) Systems, vol. 23, no. 4, pp. 609-618, 2015.

4. P. Gogna et al., "Quaternary Logic and Applications Using Multiple Quantum Well Based SWSFETs." International Journal of VLSI Design & Communication Systems VLSICS, vol. 3, no. 5, pp. 27-42, 2012.

5. S. Borisov and A. S. Korotkov. "Procedure for Building a MOS Transistor High Frequency Small-signal Model." Radio electronics and Communications Systems, vol. 53, no. 7, pp. 356-66, 2010.

6. B. Saman, P. Mirdha, M. Lingalugari, P. Gogna, F. C. Jain, El-Sayed Hasaneen, and E. Heller. "Logic Gates Design and Simulation Using Spatial Wave-function Switched (SWS) FETs." International Journal of High Speed Electronics and Systems, vol. 24, no. 03n04, 2015.

7. Cadence SPICE Reference Manual, 20 Nov. 2015.

8. S-D. Kim, C-M. Park, and J. Woo. "Advanced Model and Analysis of Series Resistance for CMOS Scaling into Nanometer Regime. II. Quantitative Analysis." IEEE Trans. Electron Devices, vol. 49, no. 3, pp. 467-72, 2002.

9. M. A. Cirit, "The Meyer Model Revisited: Why Is Charge Not Conserved? (MOS Transistor)." IEEE Transactions on Computer-Aided Design of Integrated Circuits and Systems, vol. 8, no. 10, pp. 1033-037, 1989.

10. T. Zaki, Susanne Scheinert, Ingo Horselmann, Reinhold Rodel, Florian Letzkus, Harald Richter, Ute Zschieschang, Hagen Klauk, and Joachim N. Burghartz. "Accurate Capacitance Modeling and Characterization of Organic Thin-Film Transistors." IEEE Trans. Electron Devices, vol. 61, no. 1, pp. 98-104, 2014.

11. P. E. Allen and D. R. Holberg, CMOS analog circuit design, 3rd Ed. New York: Oxford University Press, 2011.

12. P. Gogna, E. Suarez, M. Lingalugari, J. Chandy, E. Heller, E.-S. Hasaneen, and F.-C. Jain. "Ge-ZnSSe Spatial Wave-function Switched (SWS) FETs to Implement Multibit SRAMs and Novel Quaternary Logic." Journal of Electronic Materials, vol. 42, no. 11, 2013.

13. B. Saman, P. Gogna, E-S. Hasaneen, E. Heller and F.C. Jain. "Spatial Wave-function Switched (SWS) FETs SRAM Circuits and Simulation." Submitted to International Journal for High Speed Electronics and Systems (IJHSES), vol. 9, 2016.

14. S. Karmakar, "Design of Three Bit Analog-To-Digital Converter (ADC) Using Spatial Wave-function Switched (SWS) FETS." International Journal of VLSI Design & Communication Systems VLSICS, vol. 4, no. 3, pp. 1-14, 2013.

15. S. Karmakar and F. C. Jain, "Design of Three bit ADC and DAC Using Spatial Wave-function Switched SWSFETs," Silicon, vol. 8, no. 3, pp. 369-379, 2016.

16. A. Bazigos, M. Bucher, F. Krummenacher, J.-M. Sallese, A.-S. Roy, C. Enz, EKV3 MOSFET Compact Model Documentation, Model Version 301.02, Technical Report, Technical University of Crete, 16 July 2008.
17. S. Yoshitomi, A. Bazigos, M. Bucher, "EKV3 Parameter Extraction and Characterization of 90nm RF-CMOS Technology" 14th Int. Conf. on Mixed Design (MIXDES 2007), Ciechocinek, 21-23 June 2007.
18. S. Yoshitomi, "Challenges of Compact Modeling for Deep-Submicron RF-CMOS Devices" 12th Int. Conf. on Mixed Design (MIXDES 2005), Krakow, Poland, 22-25 June 2005.
19. A. Bazigos, M. Bucher, EKV3.0 Model Code, Parameters & Case Studies, Presentation at EKV Model Users' Group Meeting and Workshop, November 4-5 2004, EPFL, Lausanne.
20. C. C. Enz, F. Krummenacher, and E. A. Vittoz, "An analytical mos transistor model valid in all regions of operation and dedicated to low voltage and low-current applications," Analog integrated circuits and signal processing, vol. 8, no. 1, pp. 83-114, 1995.
21. Vasundara, P.K.S, and K.S Gurumurthy. "Quaternary CMOS Combinational Logic Circuits." International Conference on Information and Multimedia Technology, 538-42, 2009.
22. Shweta Hajare and Pravin Dakhole. "Design of Quaternary Logical Circuit Using Voltage and Current Mode Logic." International Journal of VLSI Design & Communication Systems (VLSICS) vol. 8, no. 4, August 2017.
23. C. W. Yeung et al., Channel Geometry Impact and Narrow Sheet Effect of Stacked Nanosheet, IEDM, pp. 652-655, December 2018.

Low Pass Filter PUF: Authentication of Printed Circuit Boards Based on Resistor and Capacitor Variations

Shahed Enamul Quadir* and John A. Chandy[†]

*Department of Electrical and Computer Engineering, University of Connecticut,
371 Fairfield Way, Storrs, Connecticut, USA
*md.enamul_quadir@uconn.edu
[†]john.chandy@uconn.edu

Physical Unclonable Functions (PUFs) are probabilistic circuit primitives that extract randomness from the physical characteristics of a device. PUFs are easy and simple to implement and its random nature makes its behavior hard to predict and model. Most existing PUF designs are based on variation at the chip level and can not be implemented in a printed circuit board (PCB). Therefore, these PUFs can not be used to protect against counterfeit PCBs in a distributed supply chain. In this work, we propose a novel PUF design based on resistor and capacitor variations for low pass filters (LoPUF). We demonstrate the setup in a protoboard for different resistor-capacitor pairs (RC pairs) for reliable low pass filter PUF. Because of process variations, the voltage will be different at the same cut-off frequency for our proposed PUF. Finally, the output of the filter is connected to an inverter to measure the pulse width and best suitable pulses are used for ID generation based on our algorithm.

Keywords: Authentication; hardware security; printed circuit board (PCB); physically unclonable function; counterfeiting; low pass filter.

1. Introduction

Due to globalization and a diversified supply chain, counterfeit electronic products have become a major issue over the past decade[1]. Those These concerns consist of: 1) malicious modification, 2) destructive reverse engineering to get the critical information and 3) cloning and replication. A survey of intellectual property (IP) infringement published by Semiconductor Equipment and Materials International (SEMI) in 2010[2,3] showed that 90% of companies have experienced infringement for products and among them 54% have faced severe infringement. IHS Technology estimated that the risk because of counterfeit products is over \$169 billion a year[4]. Counterfeit poses a great threat to national security also. Approximately 15% of spare and replacement semiconductors purchased by the Pentagon are counterfeit[5]. The counterfeit electronic product could be sold as authentic one but it would have inconsistency in performance, functionality and reliability[6,7]. Researchers are working on for prevention, mitigation, and identification of chip-level counterfeiting. Printed Circuit Boards (PCBs) also have a diversified supply chain because they are a basic component of electronic systems. Hence, the supply chain could have multiple untrusted parties, for example, trading partners, distributors, and retailers. Therefore, it is possible for a number of different forms of counterfeiting attacks on PCBs. The most prevalent attack is cloning because the PCB can be easily counterfeited by reverse engineering. If a critical system consists of counterfeit product, it could cause serious degradation of performance and threat of security[8].

These counterfeited PCBs can lead to failure of a system or theft of sensitive information. Therefore, anti-counterfeiting solutions for PCBs is an emerging concern for electronic system security and privacy to researchers as well as for industry[7]. However, there is not enough research conducted for identification or authentication of counterfeit PCBs.

Very few effective PCB detection methodologies have been described in the literature. One method used by Applied DNA Sciences named as "DNA marking"[9]. Unique and unclonable botanical DNA are used in this method to verify the product. Laser reader could be used for authentication on the supply chain. Refs. 10 and 11 are commercial solutions which depend on the dedicated secure chip and this chip need to be integrated into to the PCB. But, those methodologies only can detect the individual chips on the PCB. Hence, this technology is only able to detect the soldered ICs but the PCB can not be verified. Radio Frequency Identification (RFID) technology can also used for verification of electronic products[12]. This method is wireless and radio-frequency electromagnetic fields are used for non-contact transfer of data. Therefore, tracking tags attached to the object could be used for identification of the object. However, RFID can also be cloned easily, meaning that the cloned RFID and authentic RFID are hard to distinguish.

Physical Unclonable Functions (PUF) have emerged as a mechanism for authentication of electronic parts[13,14]. A PUF uses random manufacturing process variability to extract a unique signature from each production unit. A PUF is a multiple-input multiple-output function with very hard to predict outputs. A set of outputs are mapped from the PUF's input and those output responses are described as Challenge-Response Pairs (CRPs) (see Fig. 1). It is very hard to predict the responses to a particular challenge because the PUF is derived from the random variation of the process. Therefore, a PUF has excellent resilience for authentication and verification. In addition, PUFs have been used for promising security measures because of its ability to generate unique signatures. For example, PUFs have been used for authentication of the device and generation of ID[13]. Figure 1 shows that system designer authenticates the particular chip for a given challenge and those challenge response pair will be recorded in the in the system. Then, the response could be verified using CRPs from the trusted data base. A number of existing PUF designs have been proposed included those based on ring oscillators[13], arbiters[15], gate glitches[16], memory arrays[17,18,19], scan chains[20], etc.

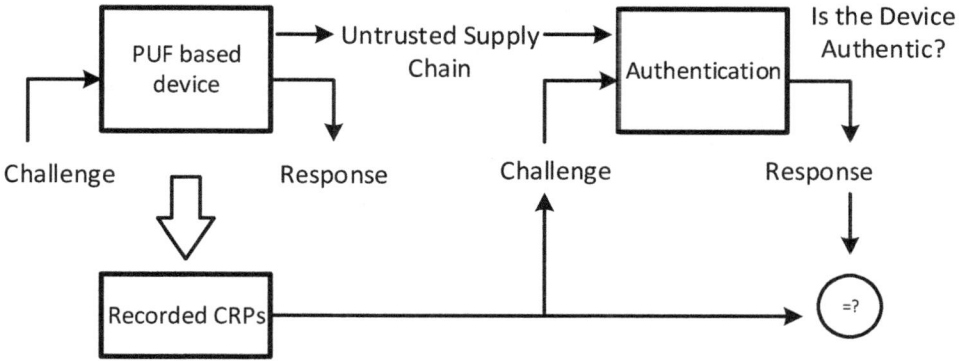

Fig. 1. Physical unclonable function.

Furthermore, PUF design has two major categories which could be mentioned as "strong PUFs" and "weak PUFs". Weak PUFs can directly digitize a "fingerprint" of the circuit and those digital signatures could be integrated into cryptographic primitives. The "fingerprint" (CRPs) should be stored in a database at enrollment. The end user can verify the authenticity of the product by checking the CRPs from the database. Therefore, PUFs should be unique, random and reliable[21]. The PUF should have a unique response and could produce a random signature in different environmental conditions.

To date, however, most of these prior PUF designs have been embedded into an integrated circuit die or chip. Therefore, these methods can not be implemented on a PCB. Because the variation sources mentioned due to intrinsic variability with the chip, not a variation from the PCB. Therefore, authentication needs to be proposed which is applicable for PCB level.

Therefore, authentication needs to be derived from manufacturing variations from PCB for security standpoint. Those variations should be able to generate an ID for differentiation between counterfeit and authenticate PCB. Few techniques have been proposed in the literature recently for PCB identification. The trace impedance variations are measured using dedicated testing equipment in Ref. 7 and used for counterfeit PCB detection. The main assumption of this authentication method attacks is that PCB is vulnerable to counterfeiting from PCB designers to system designers. This method could be used for effective PCB authentication. But, the methodology is difficult to implement and after deployment counterfeit attack has not considered. Another method has been proposed in Ref. 22 which is based on manufacturing variations of capacitive copper patterns of fabricated PCB. However, the complexity of the design make very hard to implement[22]. Therefore, in this paper, we propose a novel counterfeit PCB authentication approach that utilizes the RC pair variations on a manufactured PCB to create a unique signature. We have also shown experimental results to show the uniqueness and robustness

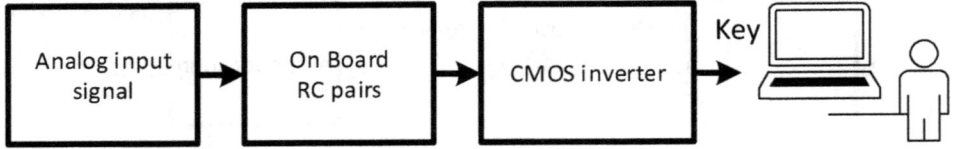

Fig. 2. Block diagram of proposed LoPUF key generation.

of our proposed PUF. It is also worth mentioning that our proposed ID generation technique is based on fully digital solution that does not require any analog circuitry to process the input analog signals from the PCB RC circuit.

The rest of our paper is organized as follows. The proposed methodology for authentication of PCB is shown in Sec. 2. The experimental results and analysis are provided in Sec. 3. Finally, in Sec. 4, we conclude and discuss future works.

2. Proposed Methodology

Our proposed method is based on the variability of resistor-capacitor (RC) pairs on a PCB. The PCB is populated with a number of RC pairs configured as a low-pass filter as shown in Fig. 3. Those RC pairs should be identical, but manufacturing variability causes slight physical variations in each pair. By applying a sinusoidal input with a frequency near the cut-off frequency of the low-pass filter, we should be able to measure variations in the response of the filter because of the variability in the RC pairs. These filter responses can be used to serve as the unique response of the LoPUF. The detailed description will be presented below.

2.1. *Low Pass RC filter*

If one had an ideal low-pass filter, one could extract a discriminating bit from the filter such that if the signal passes, we get a 1 and if not we get a 0. However, as shown in Fig. 4, real filters do not have such sharp cutoff features. One could use higher-order filters to achieve sharper cutoffs, but the complexity of these circuits make them impractical to implement. In our proposed method, we have used a passive low pass filter consisting of a series RC (Resistor-Capacitor) circuit as shown in Fig. 3. In this type of passive filter arrangement, an input signal (Vin) is applied to the RC pair and output (Vout) is measured across the capacitor. The basic characteristic of a low pass filter is that it will only allow signals from 0 Hz to the cut-off frequency, ω_c point, but will block any higher frequency. The transfer function for ideal and practical low-pass filter is shown in Fig. 4 ($K = 1$ for first order filter). For the RC low-pass filter, the attenuation of the filter as shown in Fig. 4 is

$$\frac{V_{out}}{V_{in}} = \frac{1}{\sqrt{(RC\omega)^2 + 1}} \tag{1}$$

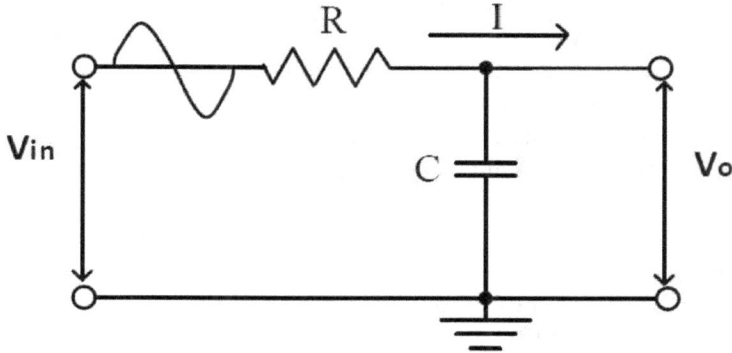

Fig. 3. Passive low pass filter.

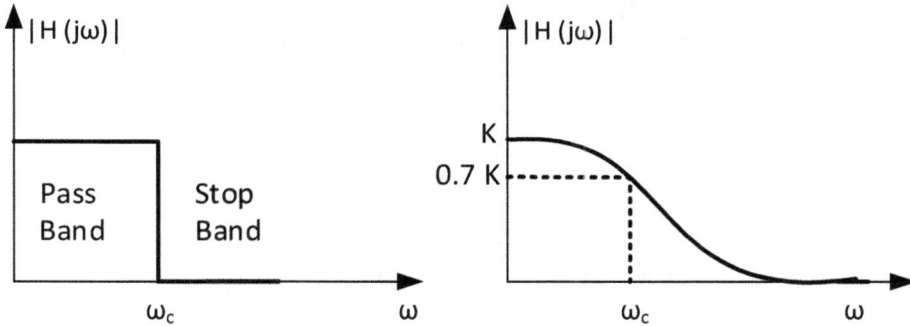

Fig. 4. Transfer characteristics of Low pass filter(Ideal and Practical).

The cut-off frequency of the filter defined as the 3dB point or the half-power frequency where the attenuation reaches $\frac{1}{\sqrt{2}}$. Thus, for the low-pass filter, the cutoff frequency is $\frac{1}{RC}$ radians/s or $\frac{1}{2\pi RC}$ Hz.

2.2. *Extraction of PUF Bits from the RC Filter*

In order to use this RC filter, we need to extract the random variations in the R and C values to create usable 1/0 bits to create a unique identifier. A simple approach is to apply a sinusoidal input to the RC filter, measure the voltage output, and use that output as our unique ID. As R and C change due to the manufacturing variations, the output will also change as well. The problem with this approach is that, assuming that the R and C variations are normally distributed, the output values will show clustering around the nominal value as described by Eq. 1. For a true unique identifier, we need the output values to be uniformly distributed. A solution to the clustered distribution problem is to instead generate a single 0/1 bit based on whether the output is greater or lesser than the nominal value. With this approach, we would need multiple RC filters to generate multiple identifier bits. Since the RC pairs can be distributed across the PCB using micro embedded components, the number of RC pairs is not a serious problem.

A more serious problem is that the design as presented requires a number of analog components including a frequency generator to create the input sinusoidal wave and an A/D converter and possible amplifier to measure the output voltage. In many embedded systems designs, adding this extra analog circuitry is not feasible. Thus, our design solution needs to be completely digital (other than the actual RC pairs on the PCB).

An inverter is connected to the output of the RC filter for our proposed method. Instead of using an analog input, we use a square wave input that can be easily generated in any digital circuit. After passing the square wave through the RC filter, we, end up with a series of exponential responses for each square wave pulse as shown in Fig. 6. In these simulations, the input square wave input is 100 Hz, and the R and C values are 47K and 47nF respectively for a cutoff frequency of 72 Hz.

Given this exponential RC filter output, we still need to extract digital bits. As mentioned above, using an A/D to measure the peak voltage of the output is not feasible. Instead of an A/D, we use an inverter at the output. As the analog output passes through the inverter, it gets converted to a series of digital pulses as shown in Fig. 7. The threshold of the inverter is 2.5V. Thus, whenever the RC output is above 2.5V, the digital output is a 0, and 1 otherwise. Variations in R and C cause changes in the length of the 0 pulse. As a result, instead of looking at the RC output voltage, we instead look at the length of the 0 pulse (which can be easily measured with a digital counter) as a means to extract the RC variations. Similar to what we discussed before, we can generate 0/1 bits by comparing the output against a nominal value.

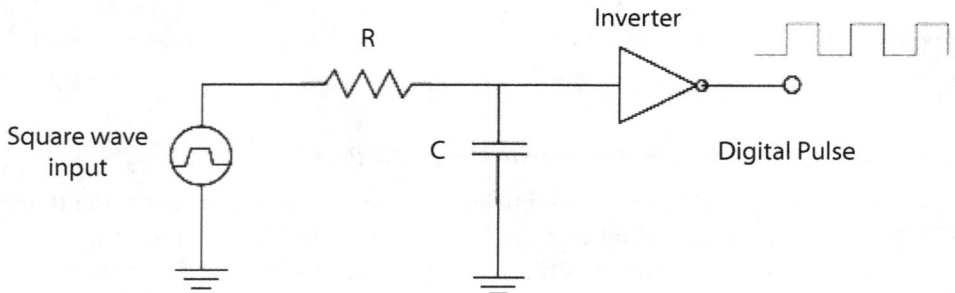

Fig. 5. Schematic diagram of our proposed method with the inverter.

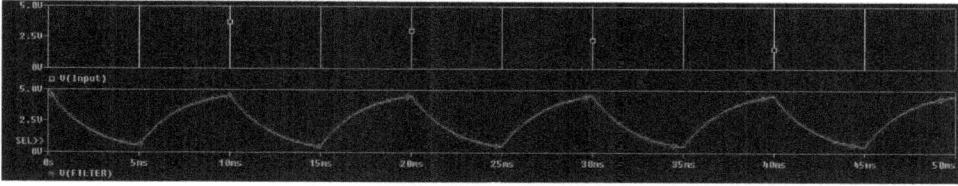

Fig. 6. Output of the RC filter.

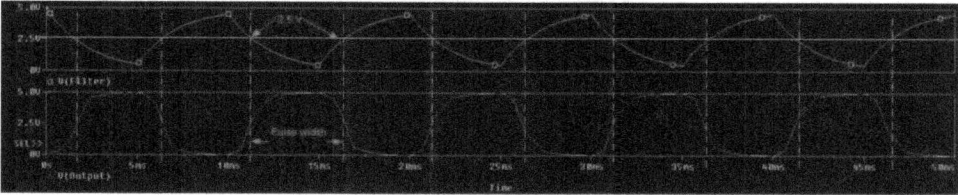

Fig. 7. Pulse width from the output of the inverter.

2.3. *Simulation Results*

We have performed simulations to check the output voltage distribution of the RC pairs. The resistors and capacitors are chosen for simulation so that it could be also applicable to the experiment. The nominal values of resistor and capacitor are used 47KΩ and 47nF respectively. The cut-off frequency for this arrangement of the filter is 72 Hz. The input voltage is chosen 1V for the simulation. The voltage distribution of the low pass filter is shown in Fig. 8 respectively using MATLAB simulation. From Fig. 8, we can see that the voltage separation between RC pairs is quite small, it will not be differentiable to generate key bits. As we mentioned in the previous section, instead we use an inverter after the output of RC pairs to convert the input signal into fully differentiable digital signal which is discussed elaborately in later sections.

3. Results and Analysis

3.1. *Experimental Setup*

We will describe experimental setup and results for the data collection of the proposed PUF in this section. The RC pairs are mounted on and wired to a protoboard and an inverter is connected after the RC filter (see Fig. 5). We have also used an Arduino Uno which is programmed to measure the pulse width from the output of the inverter. Note that in an actual implementation, this pulse width measurement could also be done with a digital counter circuit.

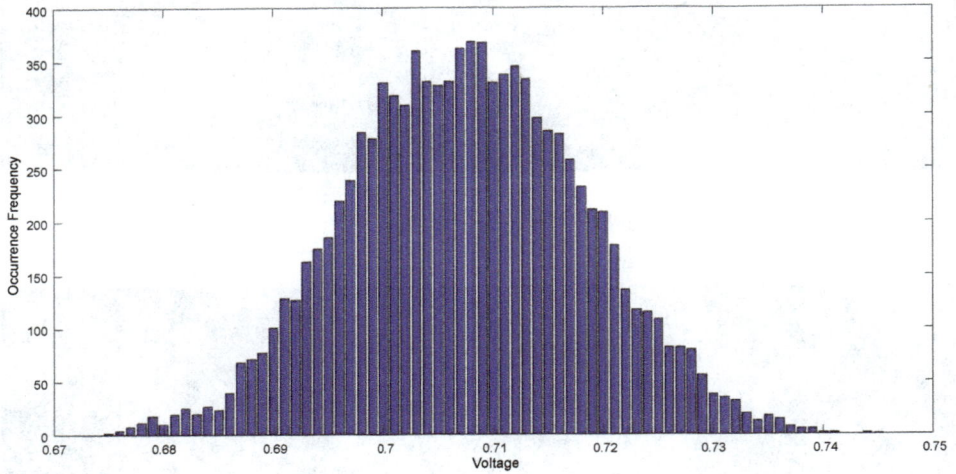

Fig. 8. Output voltage distribution of low pass filters from the MATLAB simulation.

3.2. Experimental Results

For evaluation of our proposed approach, we have used 8 resistors and 8 capacitors. As it mentioned before, the resistor and capacitor are used for the experiment 47KΩ and 47nF respectively. Therefore, for our experiment, we will have total 64 RC pair combinations. A CD4069UB CMOS Hex Inverter has been used as the inverter. The transition point of the CD4069UB is 2.5V. Therefore if the input signal less than 2.5V the output is high state and vice versa.

Finally, the output of the inverter is connected to the Arduino Uno to measure the pulse width. Note that pulse width is defined by the elapsed time between the rising and falling edges of the single pulse of a signal. A total of 200 samples of the pulse widths for each RC pair are used to get the nominal average value. In addition, the standard deviations are also recorded for every pair. The distribution of the pulse width of 64 RC pairs is shown in Fig. 9. The mean pulse width of the distribution is 8113 μs.

We fitted a distribution to our experimental data using the R statistical package *kstest* and *adtest* tools. If *kstest* and *adtest* accept the null hypothesis for our experimental data, then the data can be said to fit a normal distribution. Both tests give us "not rejected" which indicates that our data accepts the null hypothesis for normal distribution.

3.3. Pulse Width Selection Algorithm for RC Pair

The enrollment process is used to select the bits for unique ID generation of a PUF. To make the key, we have assumed that if the pulse widths are greater than 8113 μs it will be 1 and if less than 8113 μs it will be 0. However, a few pulse widths could

Histogram

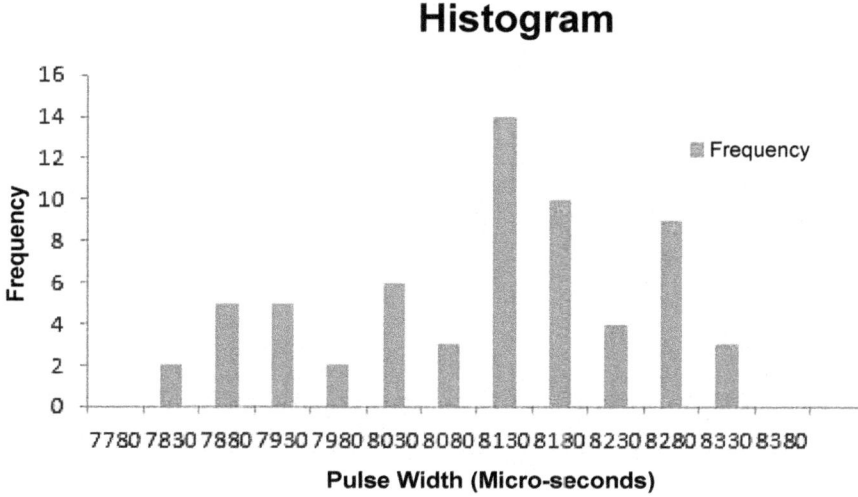

Fig. 9. Distribution of the voltages of RC pairs from the experiment.

Table 1. Total key values for different Δ, p_{eb} and acceptance regions.

Δ	1	5	10	15	20	25	30
p_{eb}	0.0318	0.0266	0.0209	0.0162	0.0123	0.0092	0.0067
Accepted region	0.994	0.971	0.944	0.916	0.887	0.860	0.833
Accepted bits (n=36)	36	37	38	39	40	41	43
Accepted bits (n=48)	48	49	50	52	54	55	57
Accepted bits (n=64)	64	65	67	69	72	74	76
Accepted bits (n=128)	128	131	135	139	144	148	153
Accepted bits (n=256)	257	263	271	279	288	297	307

be overlapping with mean pulse width 8113 μs ($\bar{\mu}$). The ideal case for RC pair key generation is that the standard deviation of the pulse widths is close to zero, but practically we got the standard deviation for the same 200 samples of the pulse width. We will define error due to overlapping is Δ. We could control Δ for how much error we could tolerate. Now, we need to define following parameters to find Δ.

$$\alpha_{left} = \int_{-\infty}^{\bar{\mu}} f(x)\,dx \tag{2}$$

Equation 2 is for error of bit 1 due to overlapping distribution to left side. and

$$\alpha_{right} = \int_{\bar{\mu}}^{+\infty} f(x)\,dx \tag{3}$$

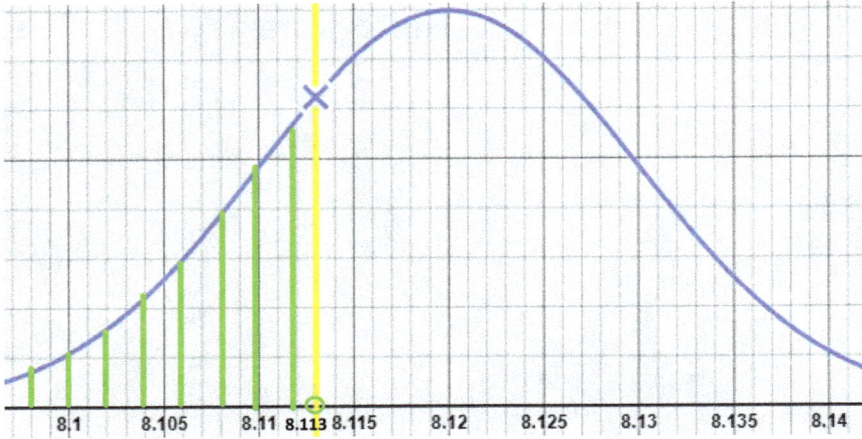

Fig. 10. An example of overlapping of pulse width which could create error bit.

Similarly, Equation 3 is for error of bit 0 due to overlapping distribution to right side.

Where, the distribution function,

$$f(x, \mu, \sigma) = \frac{1}{\sqrt{2\pi\sigma_i^2}} e^{-\frac{1}{2}\left(\frac{x-\mu_i}{\sigma_i}\right)^2} \tag{4}$$

and μ_i and σ_i is mean and standard deviation of each individual pulse width respectively (i = 1 to 64).

We have calculated the standard deviation of each pulse width and worst case standard deviation is accounted for error formulation. Our worst case standard deviation for an individual pulse width was 10 μs or 0.01 ms and Fig. 10 is shown for overlapping for a pulse width of 8120 μs or 8.12 ms as an Example.

Now, the maximum overlapping area for a single pulse width could be defined as Δ and maximum error due to overlapping,

$\Delta_{max} = 3 \times \sigma_i$ (we assumed that Δ_{max} is same in both side from the mean $\bar{\mu}$)

Therefore, the region of acceptance could be defined as

$$P_{acceptance} = 1 - \int_{\bar{\mu}-\Delta}^{\bar{\mu}+\Delta} F(x)\, dx \tag{5}$$

Where F(x) is the distribution of 64 sample pulse widths. If the pulse widths are not in the accepted region, it will be discarded. Then, we will need few more RC pair to get the desired ID.

Now, we have used the following integration in this paper to find the probability of getting the error bits,

$$p_{eb} = 2 * \int_{\bar{\mu}+\Delta}^{+\infty} \int_{-\infty}^{\bar{\mu}} F(y, \bar{\mu}, \bar{\sigma}) f(x, y, \sigma_i)\, dx\, dy, \tag{6}$$

where $\bar{\sigma}$ *and* σ_i are standard deviation of 64 pulse widths and each individual pulse width respectively.

Lets define k as the ratio between Δ and σ_i

k $= \frac{\Delta}{\sigma_i}$, therefore $\Delta = $ k σ_i, where $k_{max} = 3$.

Our approach shows that different Δ parameter and constraints will give different acceptance region and therefore different extra RC pairs to get the same desired key value.

If we get m correct bits out of n bits, the binomial probability density function for a given pair of parameters n and $P_{Acceptance}$ is

$$f(m \mid n, p_{acceptance}) = \binom{n}{m} p_{acceptance}^m q^{n-m} I_{(n=0 \ to \ 64)} \tag{7}$$

where q $= 1\text{-}p_{acceptance}$. The resulting function, f(m|n, $p_{acceptance}$), is the probability of observing m correct bits from n bits from the acceptance region. The function $I_{(n=0 \ to \ 64)}$ only take integer values from 0 to 64. In Table 1, we show the total key values needed to get the expected key values for different Δ, p_{eb} and acceptance region. For lower Δ, the accepted key bits are smaller because of wide acceptance region. As delta is increased the constraints make acceptance region narrower- therefore more bits will be needed. Figure 11 shows accepted key bits when accepted region varies. For larger accepted region (lower Δ), we will need a relatively small number of accepted key bits with less reliability.

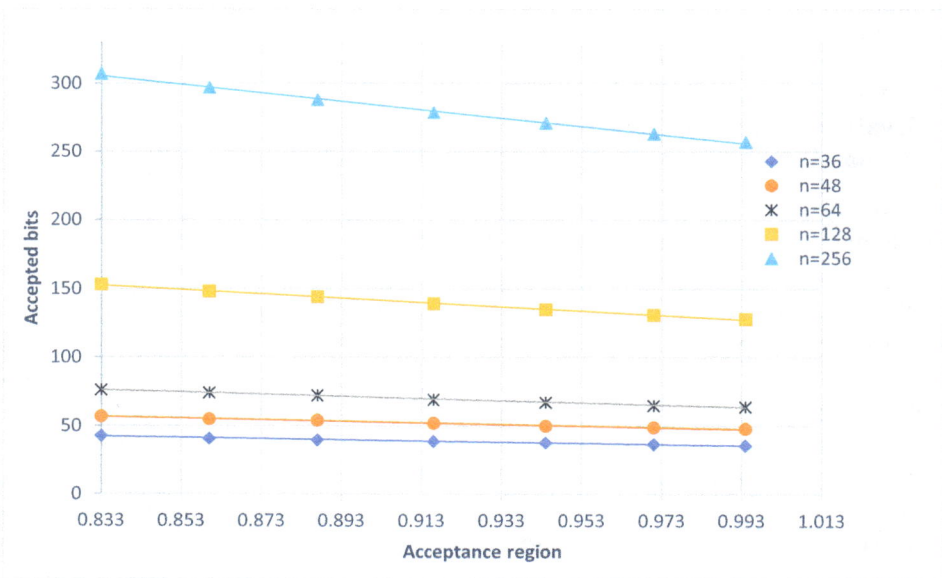

Fig. 11. Accepted bits for different acceptance region for key bits (n) of 36, 48, 64, 128 and 256.

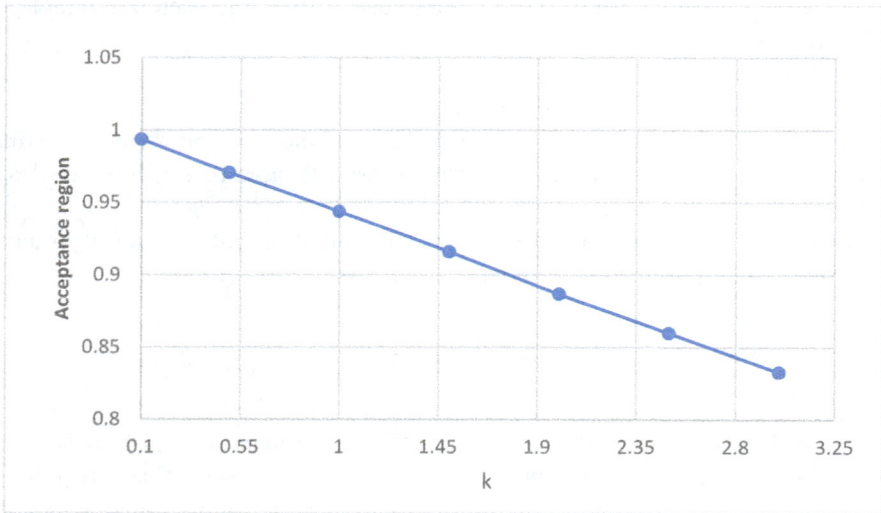

Fig. 12. Accepted region for different k values.

Figure 12 shows the acceptance regions with varying k. With increasing k, acceptance region decreases which gives less accountable key bits. However, for higher k, the constraints are stricter which gives more reliable key bits. Therefore, we need to add more RC pairs for strict constraints to get our desired key. More RC pairs will add more area in the board. Thus, constraints offer choice between area and reliability.

4. Evaluation of LoPUF Under Different V_{DD} and Temperature Variations

We have examined the pulse width stability of LoPUF under various supply voltage and temperature conditions. We examined the pulse width variation at different supply voltages (4.5V, 5.5V) and temperatures ($25°C$ to $85°C$). From the experiment, we have found that at higher supply voltage (5.5V), the pulse width increases and at lower supply voltage, the pulse width decreases. Thus, power supply variation can alter the pulse width of the LoPUF. Therefore, a controlled voltage regulator is needed to regulate the supply voltage at the nominal value (5V) for LoPUF stability. For the temperature variation, Temptronic TP04100A ThermoStream Thermal Inducting System is used from $25°C$ to $85°C$. The test setup is shown in Fig. 13. We have used temperature condition from $25°C$ to $85°C$ with $10°C$ interval, as Thermostream system could deliver controlled temperature precisely. As temperature increases, % variation of the pulse width increases, as shown in Fig. 14. Therefore, authentication of the PCB using the RC pair method for the LoPUF must be done at room temperature.

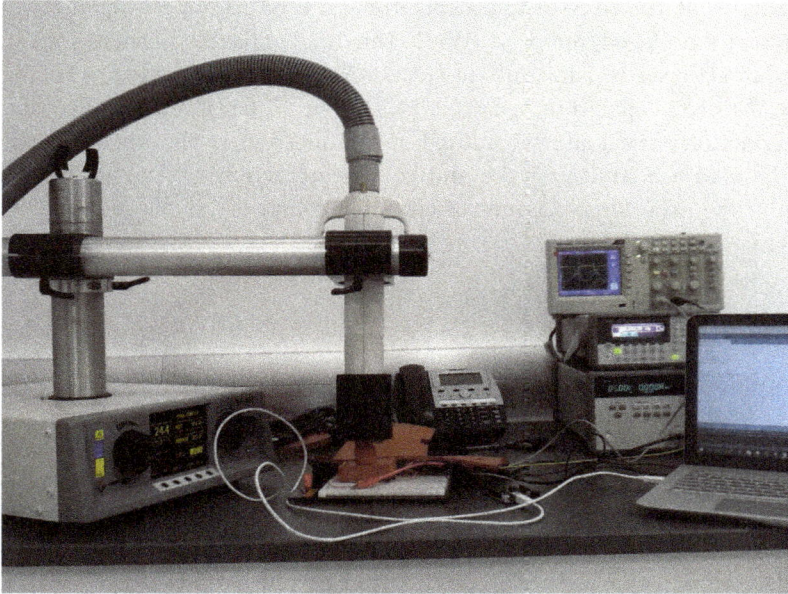

Fig. 13. Test setup for temperature variation with Thermostream.

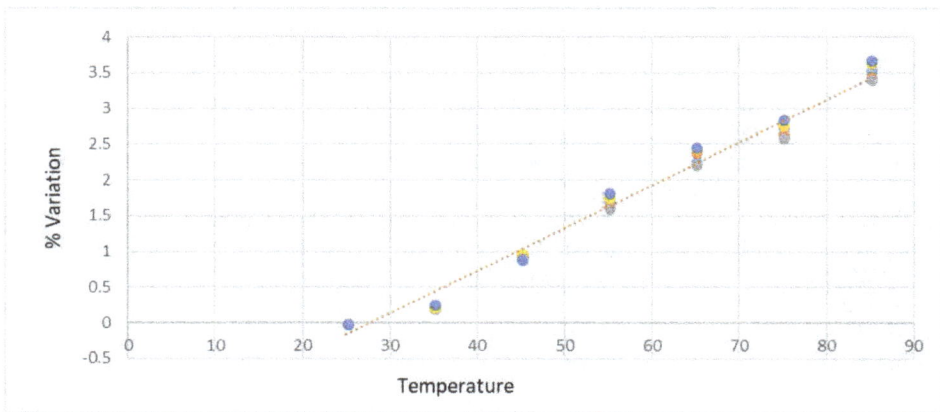

Fig. 14. % variation versus temperature for the RC pairs.

5. Security Measures of the LoPUF Against Attacks

We have presented LoPUF based on resistor and capacitor variation of low pass filter for system and board level authentication. In the literature, machine learning attacks have been reported for strong PUFs[23]. As our proposed design is a weak PUF, it is not vulnerable to machine learning attacks. Another vulnerability of the design is that the resistor and capacitor are used as discrete components that could be easily replaced/removed. While this might disable authentication, it will not allow an attacker to fake authentication. Nevertheless, to address this issue, we could use "buried" capacitors as used in BoardPUF[22]. Even if we use the BoardPUF buried capacitors, the main advantage of our method is the area efficiency. The BoardPUF uses a Schmitt trigger and comparator circuit to create the PUF ID, which can take up a large amount of circuitry compared to the minimal inverters that we use.

6. Conclusion

It is very important for a company to prevent unrecoverable losses due to counterfeiting. In this paper, we have presented a PUF based authentication technique to prevent counterfeiting. LoPUF generate key bits which are reliable and unique and those key bits are generates utilizing the manufacturing variations of resistors and capacitors. An algorithm also described to find reliable bits for enrollment. This algorithm is applied to different key values and there is a trade off between reliable key values and area overhead. Although our proposed LoPUF has many advantages, we have to consider many other challenges, for example, area overhead, power, environmental and supply voltage variations, cost etc. In future work, we will fabricate the board to investigate the performance in real situations.

References

1. M. M. Tehranipoor, U. Guin, and D. Forte, "Counterfeit integrated circuits," in *Counterfeit Integrated Circuits*. Springer, 2015, pp. 15–36.
2. A. C. Baumgarten, *Preventing integrated circuit piracy using reconfigurable logic barriers*. Iowa State University, 2009.
3. S. E. Quadir, J. Chen, D. Forte, N. Asadizanjani, S. Shahbazmohamadi, L. Wang, J. Chandy, and M. Tehranipoor, "A survey on chip to system reverse engineering," *ACM journal on emerging technologies in computing systems (JETC)*, vol. 13, no. 1, p. 6, 2016.
4. M. Tehranipoor, H. Salmani, and X. Zhang, "Counterfeit ICs: Taxonomies, assessment, and challenges," in *Integrated Circuit Authentication*. Springer, 2014, pp. 161–178.
5. P. Subramanyan, S. Ray, and S. Malik, "Evaluating the security of logic encryption algorithms," in *Hardware Oriented Security and Trust (HOST), 2015 IEEE International Symposium on*. IEEE, 2015, pp. 137–143.
6. U. Guin, K. Huang, D. DiMase, J. M. Carulli, M. Tehranipoor, and Y. Makris, "Counterfeit integrated circuits: A rising threat in the global semiconductor supply chain," *Proceedings of the IEEE*, vol. 102, no. 8, pp. 1207–1228, 2014.
7. F. Zhang, A. Hennessy, and S. Bhunia, "Robust counterfeit PCB detection exploiting intrinsic trace impedance variations," in *VLSI Test Symposium (VTS), 2015 IEEE 33rd*. IEEE, 2015, pp. 1–6.
8. S. Ghosh, A. Basak, and S. Bhunia, "How secure are printed circuit boards against trojan attacks?" *IEEE Design & Test*, vol. 32, no. 2, pp. 7–16, 2015.

9. J. A. Hayward and J. Meraglia, "DNA marking and authentication: A unique, secure anti-counterfeiting program for the electronics industry," in *International Symposium on Microelectronics*, vol. 2011, no. 1. International Microelectronics Assembly and Packaging Society, 2011, pp. 000 107–000 112.

10. Renesas, "Board ID," http://am.renesas.com/products/security/boardid/, 2014.

11. Maxim, "PCB ID and authentication," http://www.maximintegrated.com/en/products/comms/one-wire/pcb-id-and-authentication.html, 2014.

12. S. Devadas, E. Suh, S. Paral, R. Sowell, T. Ziola, and V. Khandelwal, "Design and implementation of PUF-based 'unclonable' RFID ICs for anti-counterfeiting and security applications," in *RFID, 2008 IEEE International conference on*. IEEE, 2008, pp. 58–64.

13. G. E. Suh and S. Devadas, "Physical unclonable functions for device authentication and secret key generation," in *Proceedings of the 44th annual design automation conference*. ACM, 2007, pp. 9–14.

14. M. T. Rahman, D. Forte, F. Rahman, and M. Tehranipoor, "A pair selection algorithm for robust RO-PUF against environmental variations and aging," in *Computer Design (ICCD), 2015 33rd IEEE International Conference on*. IEEE, 2015, pp. 415–418.

15. J. W. Lee, D. Lim, B. Gassend, G. E. Suh, M. Van Dijk, and S. Devadas, "A technique to build a secret key in integrated circuits for identification and authentication applications," in *VLSI Circuits, 2004. Digest of Technical Papers. 2004 Symposium on*. IEEE, 2004, pp. 176–179.

16. D. Suzuki and K. Shimizu, "The glitch PUF: A new delay-PUF architecture exploiting glitch shapes," in *International Workshop on Cryptographic Hardware and Embedded Systems*. Springer, 2010, pp. 366–382.

17. D. E. Holcomb, W. P. Burleson, and K. Fu, "Power-up SRAM state as an identifying fingerprint and source of true random numbers," *IEEE Transactions on Computers*, vol. 58, no. 9, pp. 1198–1210, 2009.

18. A. R. Krishna, S. Narasimhan, X. Wang, and S. Bhunia, "MECCA: A robust low-overhead PUF using embedded memory array," in *International Workshop on Cryptographic Hardware and Embedded Systems*. Springer, 2011, pp. 407–420.

19. F. Tehranipoor, N. Karimian, K. Xiao, and J. Chandy, "DRAM based intrinsic physical unclonable functions for system level security," in *Proceedings of the 25th edition on Great Lakes Symposium on VLSI*. ACM, 2015, pp. 15–20.

20. Y. Zheng, A. R. Krishna, and S. Bhunia, "ScanPUF: Robust ultralow-overhead PUF using scan chain," in *Design Automation Conference (ASP-DAC), 2013 18th Asia and South Pacific*. IEEE, 2013, pp. 626–631.

21. R. Maes and I. Verbauwhede, "Physically unclonable functions: A study on the state of the art and future research directions," in *Towards Hardware-Intrinsic Security*. Springer, 2010, pp. 3–37.

22. L. Wei, C. Song, Y. Liu, J. Zhang, F. Yuan, and Q. Xu, "BoardPUF: Physical unclonable functions for printed circuit board authentication," in *Computer-Aided Design (ICCAD), 2015 IEEE/ACM International Conference on*. IEEE, 2015, pp. 152–158.

23. G. Hospodar, R. Maes, and I. Verbauwhede, "Machine learning attacks on 65nm arbiter pufs: Accurate modeling poses strict bounds on usability," in *Information Forensics and Security (WIFS), 2012 IEEE International Workshop on*. IEEE, 2012, pp. 37–42.

Interaction Length for Dislocations in Compositionally-Graded Heterostructures

Minglei Cai[1], Tedi Kujofsa[2], Xinkang Chen[3], Md Tanvirul Islam[4] and John E. Ayers[5]

Electrical and Computer Engineering Department,
371 Fairfield Way, Unit 4157, Storrs, CT 06269-4157, USA
[1]*minglei.cai@uconn.com*
[2]*tedi.kujofsa@gmail.com*
[3]*xinkang.chen@uconn.edu*
[4]*md.t.islam@uconn.edu*
[5]*john.ayers@uconn.edu*

Several simple models have been developed for the threading dislocation behavior in heteroepitaxial semiconductor materials. Tachikawa and Yamaguchi [*Appl. Phys. Lett., 56, 484 (1990)*] and Romanov *et al.* [*Appl. Phys. Lett., 69, 3342 (1996)*] described models for the annihilation and coalescence of threading dislocations in uniform-composition layers, and Kujofsa *et al.* [*J. Electron. Mater., 41, 2993 (2013)*] extended the annihilation and coalescence model to compositionally-graded and multilayered structures by including the misfit dislocation-threading dislocation interactions. However, an important limitation of these previous models is that they involve empirical parameters. The goal of this work is to develop a predictive model for annihilation and coalescence of threading dislocations which is based on the dislocation interaction length L_{int}. In the first case if only in-plane glide is considered the interaction length is equal to the length of misfit dislocation segments while in the second case glide and climb are considered and the interaction length is a function of the distance from the interface, the length of misfit dislocations, and the density of the misfit dislocations. In either case the interaction length may be calculated using a model for dislocation flow. Knowledge of the dislocation interaction length allows predictive calculations of the threading dislocation densities in metamorphic device structures and is of great practical importance. Here we demonstrate the latter model based on glide and climb. Future work should compare the two models to determine which is more relevant to typical device heterostructures.

Keywords: Dislocation interaction length; threading dislocations; semiconductor heterostructures and devices.

1. Introduction

Several simple models have been developed for the threading dislocation behavior in heteroepitaxial semiconductor materials. Tachikawa and Yamaguchi [1] and Romanov *et al.* [2] described models for the annihilation and coalescence of threading dislocations in uniform-composition layers. If only second-order dislocation reactions are included, the dependence of the dislocation density D with distance from the interface z is found by

$$\frac{dD}{dz} = -CD^2 \tag{1}$$

where C is a constant which must be determined empirically. However, this semi-empirical model is not predictive, and only applies to uniform-composition layers. Kujofsa et al. [3] extended the annihilation and coalescence model to compositionally-graded and multi-layered structures by including the misfit dislocation-threading dislocation interactions. This model is given by

$$\frac{dD}{dz} = \frac{4\rho_A(z)}{L_{MD}(z)sign \int_0^z \rho_A(\xi)d\xi} - CD^2 \tag{2}$$

where $\rho_A(z)$ is the areal density of misfit dislocations and $L_{MD}(z)$ is the length of misfit dislocations.

2. Dislocation Interaction Length Model

Here we present a predictive model for annihilation and coalescence of threading dislocations which is based on the dislocation interaction length $L_{int}(z)$, which determines the constant for second-order interactions:

$$\frac{dD}{dz} = \frac{4\rho_A(z)}{L_{MD}(z)sign \int_0^z \rho_A(\xi)d\xi} - L_{int}(z)D^2 \tag{3}$$

In the first case for which only in-plane glide is considered the interaction length is given by the length of misfit dislocation segments $L_{MD}(z)$, but in the second case both glide and climb are considered and the interaction length depends on the length of misfit dislocations $L_{MD}(z)$ and the distance to the closest misfit dislocation $q(z)$, and it is limited by the smaller of the two:

$$L_{int}(z) = \left(\frac{1}{L_{MD}(z)} + \frac{1}{q(z)}\right)^{-1}. \tag{4}$$

Here $q(z)$ is given by

$$q(z) = \left\{ \left[\frac{\int_0^h \rho(\xi)|z - \xi|d\xi}{\int_0^h \rho(\xi)d\xi}\right]^{-1} + \left[\frac{C}{\sqrt{\rho_{A(z)}}}\right]^{-1} \right\}^{-1} \tag{5}$$

where h is the layer thickness and C is a constant given by

$$C = \frac{\sqrt{2} + ln(1 + \sqrt{2})}{3\sqrt{2}} \approx 0.5. \tag{6}$$

In this paper we focus on the second case, but future work should compare the two models to experimental results for the purpose of determining which model is more relevant to practical device heterostructures.

3. Results and Discussion

To demonstrate the use of this model, we have calculated the interaction length in the second case for four types of compositional profiles using an assumed value of $L_{MD} = 1\,\mu m$ for the purpose of illustration. (It should be noted that in the practical application of the interaction length the depth profile of $L_{MD}(z)$ should be determined by plastic flow calculations using either a Matthews, Mader, and Light (MML) model [4] or the Dodson & Tsao / Kujofsa & Ayers (DTKA) model [5].) For a single uniform lattice-mismatched layer as shown in Fig. 1, the misfit dislocations are concentrated at the substrate interface and the interaction length therefore increases monotonically from zero. The distance to the closest misfit dislocation $q(z)$ approaches the layer thickness. For the case of a mismatched layer on top of a lattice-matched layer as shown in Fig. 2, the misfit dislocations are concentrated at the middle interface, and $q(z)$ increases monotonically from a value of zero at the middle interface. Figure 3 illustrates the case of a uniform layer on top of a graded layer, for which the misfit dislocation density is distributed uniformly throughout the first half of the epilayer. The interaction length is nearly constant in the region of uniform misfit density and then increases monotonically in the dislocation-free zone. Finally, in Fig. 4, the misfit dislocation density is assumed to be constant throughout the layer, as in a graded layer with a negligible misfit dislocation free zone. Here the interaction length is nearly constant throughout. The model performs as expected for these four well-known cases, and should allow predictions of threading dislocation densities in the more general case. This should allow optimization of graded buffer layers without the need for extensive empirical studies.

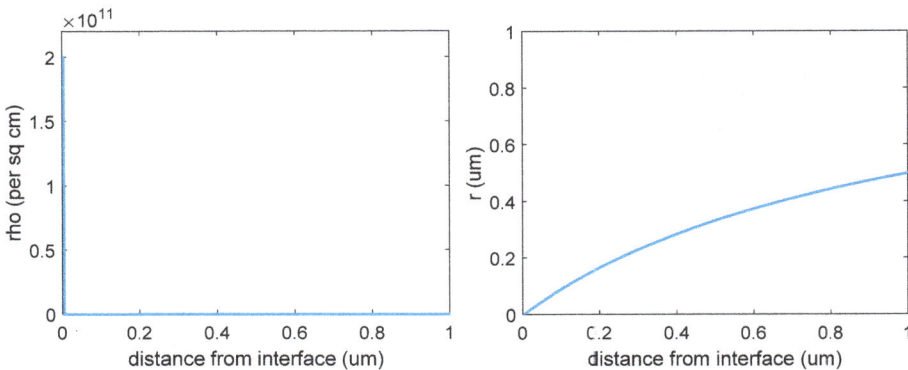

Fig. 1. Misfit dislocation density and dislocation interaction length for a mismatched uniform-composition layer with misfit dislocations at the interface.

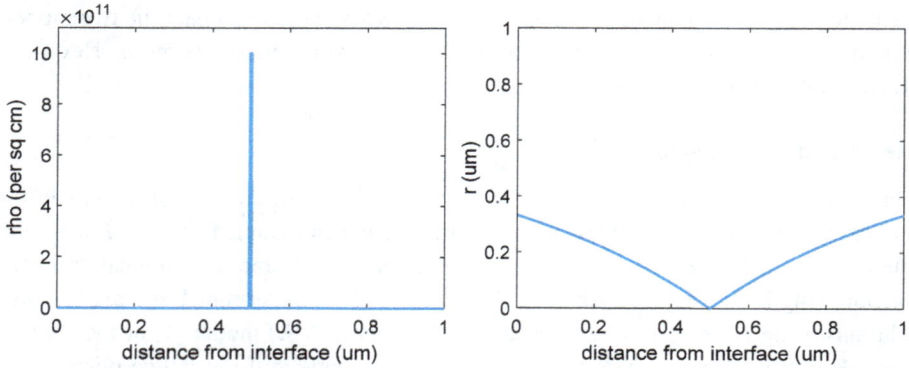

Fig. 2. Misfit dislocation density and dislocation interaction length for a mismatched uniform-composition layer on a lattice-matched layer with misfit dislocations at the middle interface located at $h_B/2$.

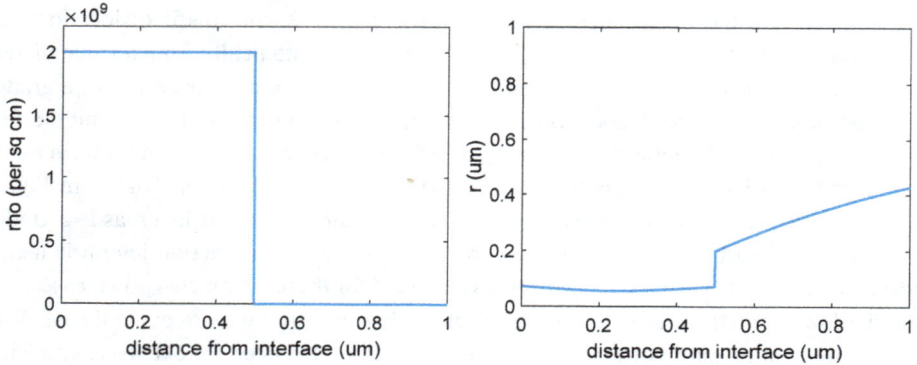

Fig. 3. Misfit dislocation density and dislocation interaction length for a graded layer with uniform misfit dislocation density and with a pseudomorphic layer on top.

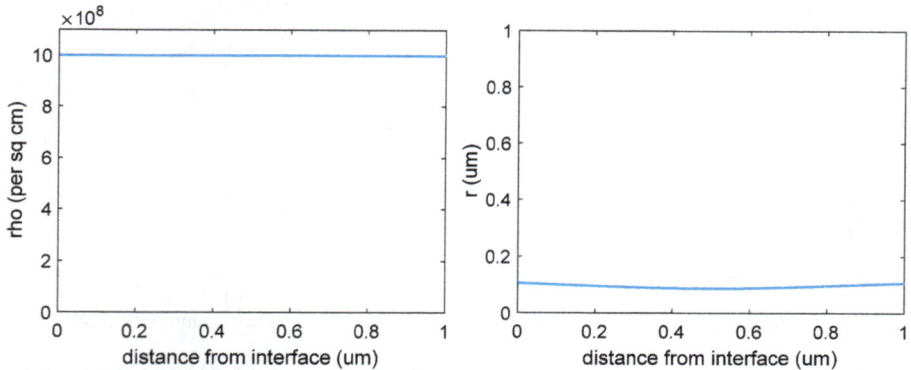

Fig. 4. Misfit dislocation density and dislocation interaction length for a graded layer with uniform misfit dislocation density throughout.

4. Conclusion

We have developed a model for the interaction length for annihilation-coalescence reactions in metamorphic semiconductor heterostructures, which includes glide and climb of dislocations. This model allows prediction of the threading dislocation density profile in a general device heterostructure.

References

[1] M. Tachikawa and Yamaguchi, *Appl. Phys. Lett.*, 56, 484 (1990).

[2] A. E. Romanov, W. Pompe, G. E. Beltz, and J. S. Speck, *Appl. Phys. Lett.*, 69, 3342 (1996).

[3] T. Kujofsa, W. Yu, S. Cheruku, B. Outlaw, S. Xhurxhi, F. Obst, D. Sidoti, B. Bertoli, P. B. Rago, E. N. Suarez, F. C. Jain and J. E. Ayers, *J. Electron. Mater.*, 41, 2993 (2013).

[4] J. W. Matthews, S. Mader, and T. B. Light, *J. Appl. Phys.*, 41, 3800 (1970).

[5] T. Kujofsa, S. Cheruku, W. Yu, B. Outlaw, S. Xhurxhi, F. Obst, D. Sidoti, B. Bertoli, P. B. Rago, E. N. Suarez, F. C. Jain, and J. E. Ayers, *J. Electron. Mater.*, 42, 2764 (2013).

Optimization of Graded Buffer Layers for Metamorphic Semiconductor Devices

Tedi Kujofsa[1], Minglei Cai[2], Xinkang Chen[3], Md Tanvirul Islam[4] and John E. Ayers[5]

Electrical and Computer Engineering Department,
371 Fairfield Way, Unit 4157, Storrs, CT 06269-4157, USA
[1]*tedi.kujofsa@gmail.com*
[2]*minglei.cai@uconn.com*
[3]*xinkang.chen@uconn.edu*
[4]*md.t.islam@uconn.edu*
[5]*john.ayers@uconn.edu*

Metamorphic semiconductor devices such as high electron mobility transistors (HEMTs), light-emitting diodes (LEDs), laser diodes, and solar cells are grown on mismatched substrates and typically exhibit a high degree of lattice relaxation. In order to minimize the incorporation of threading defects it is common to use a linearly-graded buffer layer to accommodate the mismatch between the substrate and device layers. However, some work has suggested that buffer layers with non-linear grading could offer superior performance in terms of limiting the surface density of threading defects. In this work, we have compared S-graded buffer layers with different orders and thicknesses. To do so we calculated the expected surface threading dislocation density for each buffer design assuming a GaAs (001) substrate. The threading dislocation densities were calculated using the L_{MD} model, in which the coefficient for second-order annihilation and coalescence reactions between threading dislocations is considered to be equal to the length of misfit dislocations.

Keywords: Graded buffer layers; bi-parabolic grading; linear grading; S-grading; threading dislocations; semiconductor heterostructures and devices.

1. Motivation and Theory

A drawback of metamorphic realization of semiconductor devices is the introduction of high threading dislocation densities, which can degrade carrier mobility and device performance. Therefore it is important to understand how the device design, and specifically the design of the graded buffer layer, affects the threading dislocation density. In this paper we have considered optimization of the compositional profile in S-graded buffer layers with various thicknesses by calculating the surface threading dislocation densities. In such an S-graded buffer layer the compositional indium mole fraction x is given as a function of distance from the interface z as:

$$
x = \begin{cases}
2^{\sigma-1} x_{top} \left(\dfrac{z}{h}\right)^{\sigma}; & z \leq \dfrac{h}{2} \\[2ex]
x_{top}(1 - 2^{\sigma-1}) \left(1 - \dfrac{z}{h}\right)^{\sigma}; & z \geq \dfrac{h}{2}
\end{cases}
\tag{1}
$$

where x_{top} is the indium composition at the top of the buffer layer, h is the thickness of the buffer layer, and σ is the order of the S-graded layer.

In previous work it has been shown that the threading dislocation profile for a semiconductor device heterostructure may be found by consideration of dislocation compensation as well as annihilation and coalescence of threading dislocations [1]. The threading dislocation density D is found as a function of distance from the interface by use of the differential equation

$$\frac{dD}{dz} = \frac{4\rho(z)}{L_{MD}(z)sign \int_0^z \rho(\xi)d\xi} - CD^2 \tag{2}$$

where $\rho(z)$ is the areal density of misfit dislocations and $L_{MD}(z)$ is the length of misfit dislocations.

For the implementation of an approximate model for the threading dislocation density we have assumed the misfit dislocation density profile is the same as the equilibrium one, which is a good approximation for many metamorphic heterostructures. However, the average misfit dislocation length is a non-equilibrium value which may only be determined by plastic flow calculations. Detailed calculations of this type have been made using a Dodson & Tsao / Kujofsa & Ayers (DTKA) type model [2,3], but the necessary material parameters are not available for the application of the DTKA model in all material systems. Therefore we have used the approximate Matthews, Mader, and Light (MML) model [4] to estimate L_{MD} using the average value of lattice mismatch in the structure, and we further make the approximation that L_{MD} is constant with distance from the interface. According to the MML model, the lattice relaxation δ at time t is given approximately by

$$\delta = \beta[1 - e^{-\alpha t}] \tag{3}$$

where

$$\alpha = \frac{2Gb^3\rho_0(1 + v)cos\phi cos^2\lambda D_0 exp(-U/kT)}{(1 - v)kT} \tag{4}$$

and

$$\beta = f - \epsilon_{eq} , \tag{5}$$

where G is the shear modulus, b is the length of the Burgers vector, ρ_0 is the areal misfit dislocation density at the interface, v is the Poisson ratio, ϕ is the angle between the normal to the slip plane and that direction in the interface which is perpendicular to the intersection of the glide plane and the interface, λ is the angle between the Burgers vector and that direction in the interface which is perpendicular to the intersection of the glide plane and the interface, D_0 is the diffusion coefficient for a dislocation core, U is the activation energy for dislocation glide, k is the Boltzmann constant, T is the temperature, f is the lattice mismatch strain, and ϵ_{eq} is the equilibrium strain, found using the electric circuit model.

In this work D_0 was estimated as 1.7×10^{-9} cm^2/s for In$_x$Ga$_{1-x}$As and assumed to be independent of the indium composition; the density of misfit dislocations ρ_0 was estimated as $\rho_0 = [|f|/b]^2$.

Using the MML model for lattice relaxation and the average lattice mismatch strain for each structure, the length of misfit dislocations L_{MD} was estimated by finding the effective stress τ_{eff}:

$$\tau_{eff} = \frac{2(\beta - \delta)G(1 + v)cos\phi cos\lambda}{(1 - v)}. \tag{6}$$

The effective stress was used to determine the dislocation glide velocity,

$$v = B\tau_{eff}exp(-U/kT), \tag{7}$$

where B was assumed to be $6 \times 10^{-9}cm^3 dyn^{-1}s^{-1}$, the value which we have found in our work for GaAs-based III-V semiconductors. The dislocation glide velocity was then integrated to find the approximate final length of misfit dislocations:

$$L_{MD} = \int_{h_c/g}^{h/g} \frac{2GBcos\phi cos\lambda\beta exp(-\alpha t)exp(-U/kT)}{(1 - v)}dt, \tag{8}$$

where h_c is the thickness for the onset of lattice relaxation, h is the total buffer layer thickness, and g is the growth rate, assumed to be $5\mu m/hr$ for this work. In implementing this approximate model, we assumed T = 700°C and used the material parameters for GaAs, as follows: $b = 4 \times 10^{-8}$ cm, $v = 0.31$, $G = 32 \times 10^{10}$ dyn/cm^2, $U = 1.4$ eV, $\lambda = 60°$, and $\phi = 35.26°$.

2. Results and Discussion

We calculated threading dislocation density profiles for graded In$_x$Ga$_{1-x}$As/GaAs (001) samples, each of which involved a buffer layer graded from x = 0 to x$_{top}$, with various thicknesses. Four buffer layer thicknesses were explored: 100, 200, 400 and 800. The top indium composition was set at 10, 20 and 40%. L_{MD} was found for each buffer layer thickness, and was assumed to be independent of distance from the interface. Because L_{MD} was found approximately by the MML model using the average lattice mismatch, S-graded layers with the same thickness and top composition exhibited the same approximate value of L_{MD}. Figure 1 explores the lattice mismatch, misfit dislocation density and threading dislocation density as a function of the distance from the substrate interface for InGaAs/GaAs(001) structures with a top indium composition of 20% and varying thickness for different orders of the S-graded profiles. The results of Fig. 1 show that both the epitaxial layer thickness and the order of the profile influence the dislocation dynamics. Thicker structures exhibit a thicker dislocated zone but also contain a lower peak threading dislocation density. For a given thickness and composition, an increase in the order σ results in higher dislocation densities.

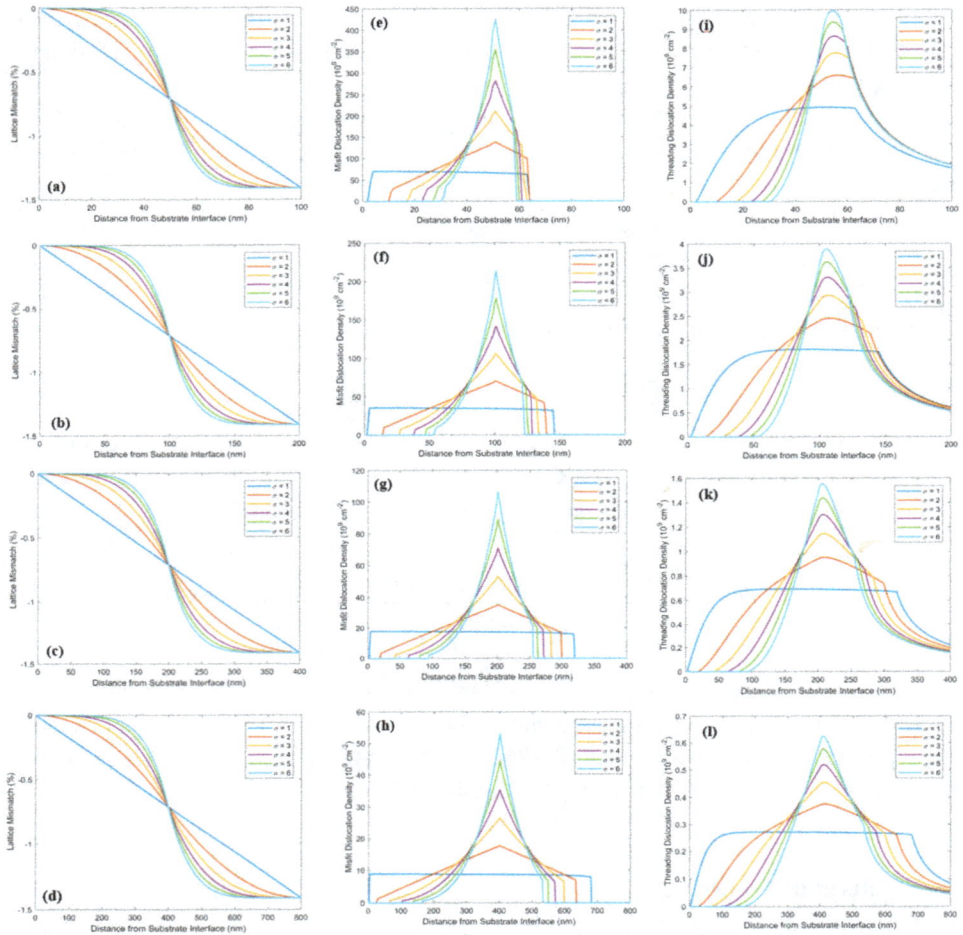

Fig. 1. (a-d) Lattice mismatch, (e-h) misfit dislocation and (i-l) threading dislocation density as a function of the distance from the interface for InGaAs/GaAs (001) structures of various thicknesses and an ending indium composition of 20%.

Figure 2 compares the surface, average and peak threading dislocation density as a function of the grading order for structures with varying thicknesses and top indium compositions. Three important trends are apparent from these results. First, for a given composition, the surface and average threading dislocation density decrease with increasing σ. Second, an increase in the epitaxial thickness results in a lower threading defect density. Third, for a given buffer layer thickness, the average (peak) TDD decreases (increases) with a higher grading order and/or top indium composition. The overall behavior for the surface TDD involves a complex interaction of these factors. However, it is generally found that the surface threading dislocation density of an S-graded layer (with $\sigma > 1$) is superior to that of a linearly-graded layer (corresponding to $\sigma = 1$).

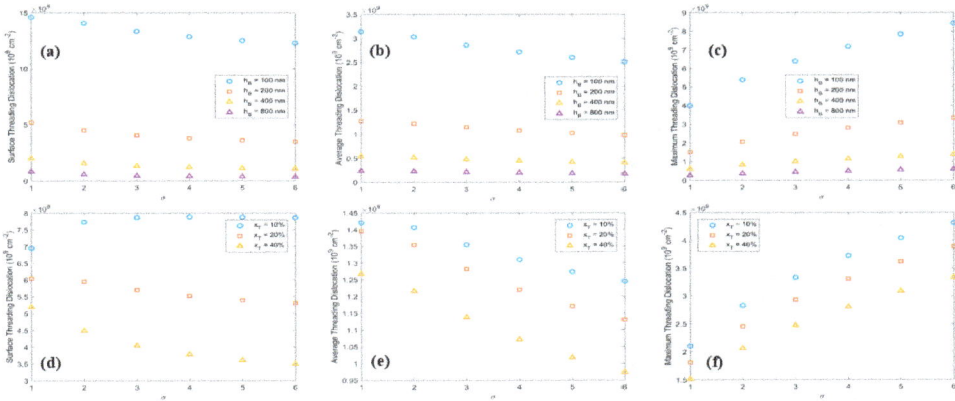

Fig. 2. Surface (a, d), average (b, e) and maximum (c, f) threading dislocation density as a function of grading order σ for (a-c) structures with a top indium composition of 40% and varying thickness as a parameter and for (d-f) 200 nm thick structures with indium composition as a parameter.

3. Conclusion

We have analyzed the threading dislocation density behavior for S-graded InGaAs/GaAs (001) buffer layers. It is generally found that the surface threading dislocation density of an S-graded layer (with $s > 1$) is superior to that of a linearly-graded layer (corresponding to $s = 1$).

References

[1] T. Kujofsa, S. Cheruku, W. Yu, B. Outlaw, S. Xhurxhi, F. Obst, D. Sidoti, B. Bertoli, P. B. Rago, E. N. Suarez, F. C. Jain, and J. E. Ayers, *J. Electron. Mater.*, 42, 2764 (2013).

[2] B. W. Dodson and J. Y. Tsao, *Appl. Phys. Lett.*, 51, 1325 (1987); *Appl. Phys. Lett.*, 52, 852 (1988).

[3] T. Kujofsa, W. Yu, S. Cheruku, B. Outlaw, S. Xhurxhi, F. Obst, D. Sidoti, B. Bertoli, P. B. Rago, E. N. Suarez, F. C. Jain and J. E. Ayers, *J. Electron. Mater.*, 41, 2993 (2013).

[4] J. W. Matthews, S. Mader, and T. B. Light, *J. Appl. Phys.*, 41, 3800 (1970).

Twin Drain Quantum Well/Quantum Dot Channel Spatial Wave-Function Switched (SWS) FETs for Multi-Valued Logic and Compact DRAMs

Husien Salama, Bandar Saman*, Evan Heller**, Raja Hari Gudlavalleti, Roman Mays, Faquir Jain#

Department of Electrical and Computer Engineering, University of Connecticut,
371 Fairfield Way, Unit 4157, Storrs, CT 06269, USA
*Department of Electrical Engineering, Taif University, Taif, KSA
**Synopsys Inc., Ossining, NY 10562, USA
#Faquir.Jain@uconn.edu

This paper aims to design and simulate a compact dynamic random access memory (DRAM) cell using two-channel spatial wavefunction switched (SWS) field-effect transistor (FET) and two capacitors. One unit of a SWSFET based DRAM cell stores 2-bits, which reduces the overall cell area by 50% as compared to a conventional 1-bit DRAM cell. SWSFETs have two or more vertically stacked quantum well channels as the transport layer between source and drain. In a two quantum channel n-SWSFET, as the gate voltage is raised above threshold, electrons appear in the lower quantum well W2 and this inversion channel connects Source S2 to drain D2. As the gate voltage is further increased, electrons transfer to upper quantum well W1 and now source S1 and drain D1 are connected electrically. Spatial location of electrons allows us to encode as 4 logic states: no electrons 00, electrons in W2 01, electrons is both wells 10 and electrons in well W1. This property of the SWSFET has been shown to implement multi-valued logic circuits. A SWSFET may have 2-4 sources and drains independently operated or connected together depending upon the logic circuit implementation.

Keywords: SWSFETs; multi-quantum well channels; DRAM; SWS-FETs, VLSI.

1. Introduction

Spatial wavefunction switched (SWS) field-effect transistors (FETs) comprise of vertically stacked quantum wells/quantum dot channels, which are selected depending upon the voltage applied on the gate. The gate controls the current flow in lower or upper quantum well channel. SWS-FET was first introduced by Jain et al. [1].

The cross-section of a SWS-FET device is shown on Fig. 1. The device has two quantum well channels as the electron transport layer between two sources and two drains. As the gate voltage is raised above the threshold, electrons appear in an n-SWSFET in the lower quantum well W2 and this inversion channel connects Source (S2) to drain (D2). As the gate voltage is increased, electrons transfer to the upper quantum well W1 and now source (S1) and drain (D1) are connected electrically. Spatial location of electrons thus provides 4 logic states: no electrons 00, electrons in lower well 01, electrons are both wells 10 and electrons in the upper well 11.

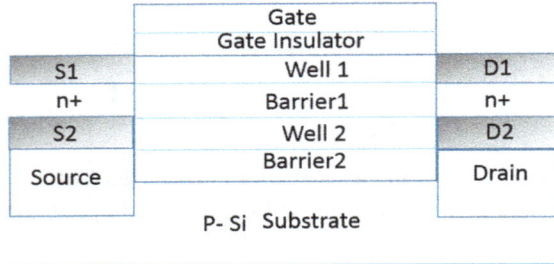

Fig. 1. The SWS-FET with Two channel.

Fig. 2 shows the energy band diagram of the two quantum wells SWS-FET type II heterostructure [2]. The Si quantum well is tensile strained where SiGe serves as a barrier. The bandgap difference is 0.15 eV (1.04-0.89) in a strained Si/Si0.5Ge0.5 heterostructure.

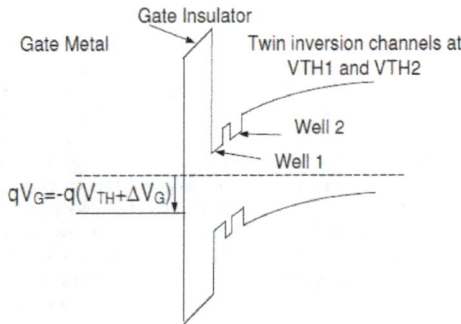

Fig. 2. The energy band diagram of two-well SWSFET structure [2].

Fig. 3(a) and Fig. 3(b) present quantum simulations showing transfer of electron wave-function from W2 to W1 as gate voltage is changed from 0.4V to 1.2V [3] with parameters listed in Table 1. The fabrication process of a SWS-FET device is compatible with conventional complementary metal-oxide-semiconductor (CMOS) FETs [2].

Table 1. Si-SiGe-TypeII-SWS Structure: (Vg=0.4, 1.2,-0.2,-0.8) [3].

Layer	Thick (um)	χ(eV)	Eg (eV)	m_e	m_h	ε_r	N_d (cm-3)	N_a (cm-3)
SiO2	0.0025	0.9	9	0.5	0.5	3.9	0.00E+00	0.00E+00
Si (QW1)	0.0025	4.15	1.04	0.19	0.49	11.9	0.00E+00	0.00E+00
SiGe(.5)	0.0015	4.0	0.89	0.13	0.38	14	0.00E+00	0.00E+00
Si (QW2)	0.0050	4.15	1.04	0.19	0.49	11.9	0.00E+00	0.00E+00
SiGe(.5)	0.0100	4.0	0.89	0.13	0.38	14	0.00E+00	0.00E+00
SiGe(.75)	0.0500	3.9	1.05	0.13	0.38	14	0.00E+00	1.00E+16
Si	0.1000	3.8	1.1	0.19	0.49	11.9	0.00E+00	1.00E+16

(a) Vg = 0.4

(b) Vg = 1.2

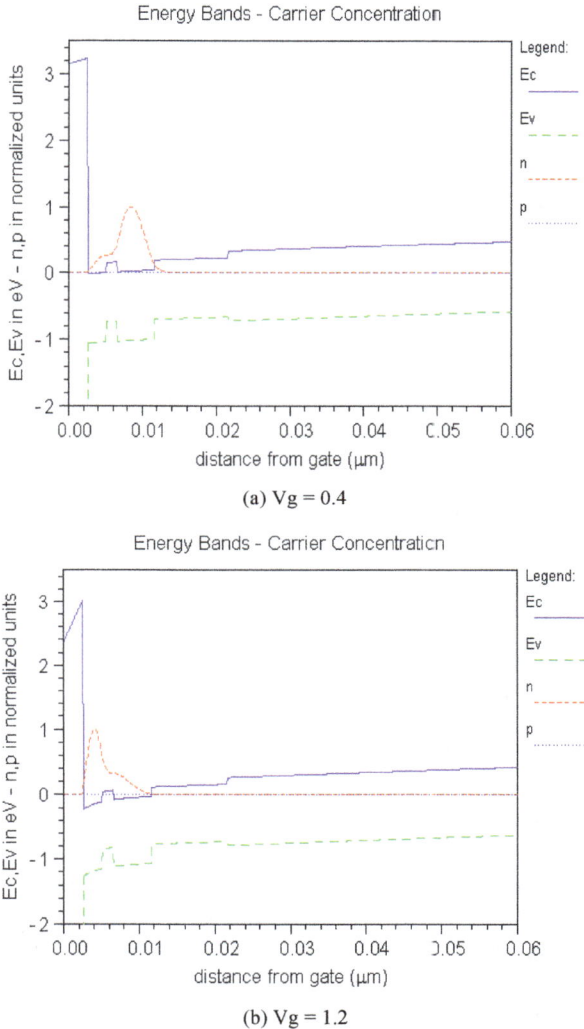

Fig. 3. Ec, Ev, n and p as a function of distance from the gate for (a) Vg=0.4, (b) Vg=1.2 [3].

The spatial switching modes of carriers are shown in Table 2. When the gate voltage (Vg) is applied between zero and below threshold voltage of well 2 (Vth2) both wells W1 & W2 are in off mode. When Vg is set above Vth2, the electrons become confined in W2 which makes the current flows in W2 (ID2). Once the gate voltage is increased and became greater than the threshold voltage of well 1 (Vth1), the electrons transfer from W2 to W1, and the current flows in W1 "ID1" as well as ID2 drop off. When Vg equals to the transition voltage (VUL), the well-2 is in off mode [4]. Present quantum simulations showing transfer of electron wavefunction from W2 to W1 as gate voltage is changed from 0.01V to 0.8V [3].

Table 2. Comparison of SWS-FETs vs. CMOS DRAM cells.

Vg (Gate Voltage)	Vth >Vg	Vth2<Vg<Vth1	Vth1<Vg<VqL	Vg>VqL
W1	0	0	1	1
W2	0	1	0	1

2. Circuit of SWSFET Modeling

The drain to source currents in a SWSFET transistor in different quantum wells, W1 (Well 1) and W2 (Well2) are governed by the gate voltage and can be given by Eq. [1-3].

$$ID = (W / L)COX\mu n((VGS - Vth)VDS - VDs^2 / 2) \tag{1}$$

$$IDS(W1) = (^W/_L)COX\mu n\left((VGS - Vth1)VDS - {VDS^2}/_2\right) \tag{2}$$

$$IDS(W2) = (W / L)COX\mu n((VGS - Vth2)VDS - (VDS^2) / 2) \tag{3}$$

The threshold voltage in (W2) can be expressed as

$$Vth\ well2\ (W2) = \{(Vth2 \qquad\qquad when\ VGSeff$$
$$< VUL@Vth2 + \alpha(VGSeff - VUL)\ when\ VGSeff > VUL)\dashv$$

Where α is the matching parameter and is given by

$$\alpha = VGS - VqL/Vq1 - Vq2$$

Here α controls the slope of the characteristics. The effective gate voltage can be expressed as $V_{GS,eff}$ = VGS- $V_{Poly,eff}$.

Where, Vth2 is the threshold voltage of well 2, Vth1 is the threshold voltage of well1, and VqL is the transition voltage. Vq1 is the voltage corresponding to peak current in well 2, α is a matching parameter. VPolyEff is the voltage drop in the Poly Si gate and VGSeff is the effective between gate-source voltages.

An inverter using a two n-channel SWS-FET is showing in Fig. 4.

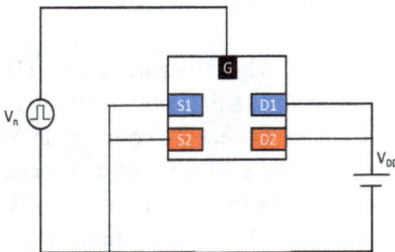

Fig. 4a. SWS-FET inverter symbol.   Fig. 4b. SWS-FET Inverter Schematic.

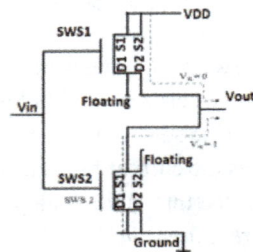

This SWS1 and SWS2 represent the symbol for a SWSFET transistor configured as an inverter. The schematic is captured in Cadence-OrCAD CIS as shown in Fig. 4. The SWSFET circuit parameters are shown in Table 3 for the implementation of inverter. The functionality of this SWS FET model can be compared to 20 nm channel length SWS FET model reported previously.

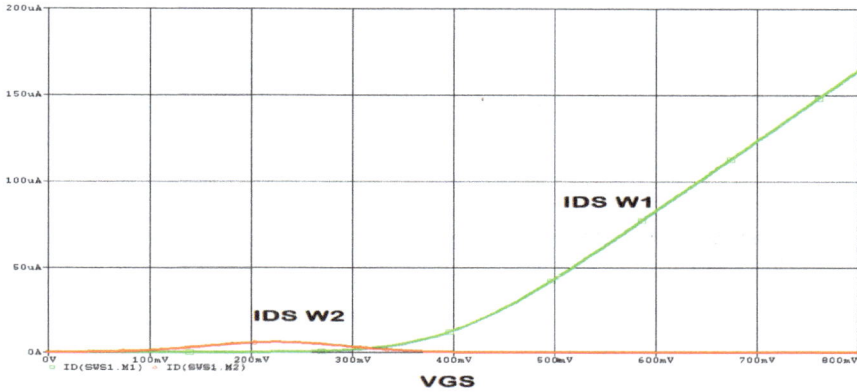

Fig. 5. SWS-FET IDS-VGS Characteristics.

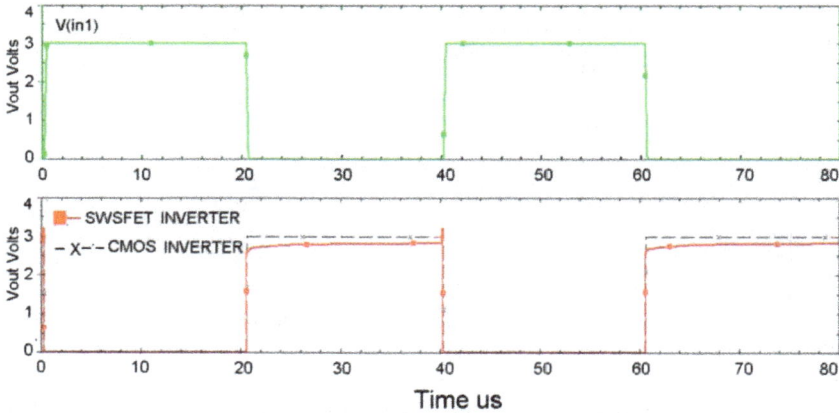

Fig. 6. Comparison of Conventional CMOS and SWS-FET Inverter input and out waveforms [7].

3. Model of DRAM Circuit

DRAM has a capacitor integrated along with the transistor. The DRAM cell stores binary information in the capacitor as electrical charges. It consists of one capacitor and one transistor as shown in Fig. 7. DRAM is implemented in a crossbar array. The capacitor C1 is charged and discharged through the transistor T1 connected between the word line (WL) and bit line (BL).

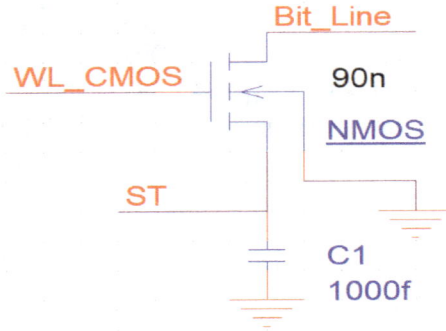

Fig. 7. Conventional DRAM unit cell.

Fig. 8 shows SWS-DRAM circuit, this circuit uses two capacitances and two bit lines to store 2-bit simultaneously.

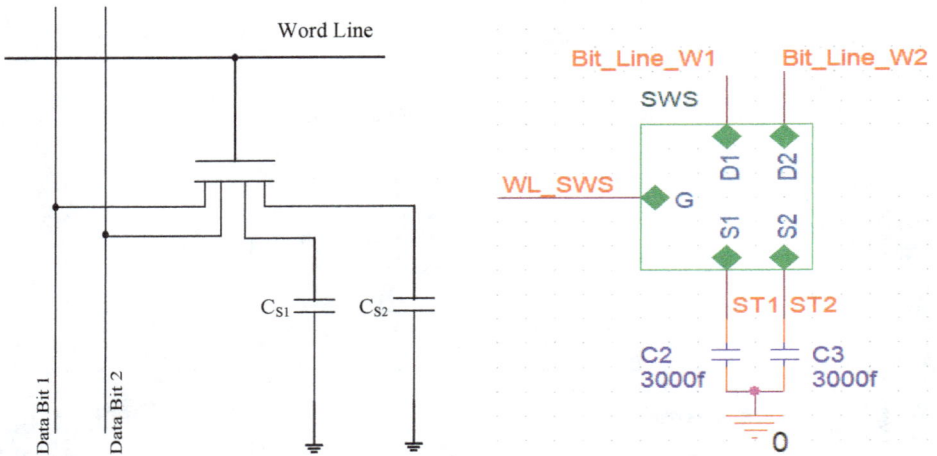

Fig. 8(a) and Fig. 8(b) shows the schematic of a two channel 2-bit DRAM cell using one SWSFET and two capacitors.

This model is set in hierarchical block and it ready to be used in Cadence-OrCAD CIS as shown in Fig. 8(a) and Fig. 8(b) shows the schematic of a two channel 2-bit DRAM cell using one SWSFET and two capacitors.

The N-MOSFET in the conventional DRAM is replaced by SWS-FET as 1T-DRAM is shown Fig. 7. There are three modes of operations of the DRAM cells.

Eraser operation: 0 Volt is applied in WL and the cell is isolated from the rest of the circuit.

Write operation: There are two approached to write required bits in the DRAM cell.
First, SWSFET acts as control the storage of multi-bit in the DRAM cell by the capacitor (C1). Similar property will store (1) in the capacitor when SWS-FET will be in (1) state.

In this way different bit value can be stored in the in the capacitor by the different state of the SWS- FET for same writing pulse time. Second, BL is biased to the voltage required to store in the capacitor. SWSFET is used as an ON or OFF switch. The bit need to be stored in the cell is applied in the BL. The BL turns ON in (VDD) voltage and WL turns ON in (VDD) which turns on the SWS. The capacitor will charge to VDD. When BL is in (0) volt, the capacitor will not charge at all.

Read operation:
BL is charged at (0) voltage. WL is turned ON to VD. The SWS-FET is ON. BL voltage will remain unchanged, if the stored charge in the capacitor is (0).

4. Simulation Results

The circuit model of a SWSFET was developed using Berkeley short channel IGFET model (BSIM 3), and the simulations are done using Cadence. In the Cadence simulator, the SWSFET channels are represented by two conventional transistors with each one having a different threshold voltage which is characteristic of a SWS-FET [10]. In this section, the performance of the proposed model is simulated, and the simulation parameters are shown in Table 3.

Table 3. Simulation Model Parameters.

Parameter	L	W	Vth1	Vth2	VUL	VDD
n-MOS	90nm	500nm	0.25 V	0.0v	0.0v	1.2v
WL-SWS-FET	90nm	500nm	0.35v	0.2v	0.25	1.2v

The truth table of the 1T DRAM is shown in Table 3. Fig. 9 shows the simulation for one SWS-FET and two capacitors (1SWS 2C) DRAM with two bit, respectively. This simulation is based on SWS-BISM model [6]. In Fig. 9, the word line (blue dotted line) selects the device, the bit line (red solid line) reads or writes the bits from/or to the capacitor (voltage on capacitor is represented by green solid line) through the device. When the word line bit goes high, the charge on bit line is transferred to the capacitor. When word line goes low, the charge on the capacitor is retrieved. The first cycle of the word line in Fig. 9 shows storing logic high and second cycle shows storing logic low on to the capacitor. For a conventional 1T-1C DRAM (top waveform in Fig. 9) shows single bit storage i.e. either logic high or low. For a SWS-FET DRAM, the charge on bit line (BL1) is stored on capacitor (CS1) if the word line voltage is greater than VTH1 (middle waveform in Fig. 9) and if the word line voltage is greater than VTH2 (< VTH1), the charge on the bit line (BL2) is stored on capacitor, CS2 (bottom waveform in Fig. 9). Here, two bits are stored on to the capacitors with a single SWSFET device depending upon the word line voltage.

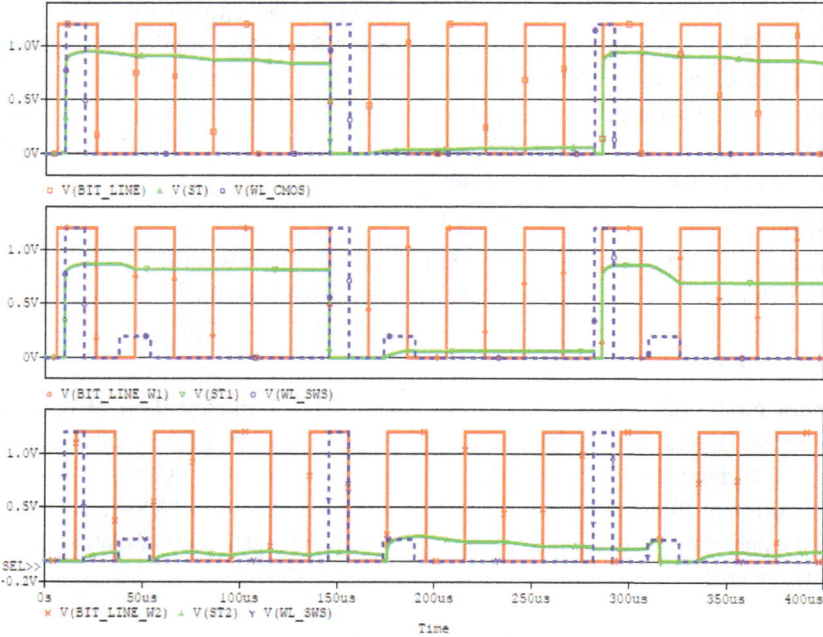

Fig. 9. The simulation of CMOS DRAM & 2Bit SWSFET DRAM.

5. Conclusion

The simulations are presented for two quantum wells (QWs) n-channel SWSFET for two bit DRAMs. As the number of devices decreases, the complexity of wiring is reduced and so is the die area. Also, it lowers the power consumption and improves the efficiency of the device. The transient simulations are presented to verify the functionality of the proposed SWS-FET DRAM with write and read operation, and compared with conventional 1T-1C DRAM. The proposed cell has achieved improved write and read ability. Even though multi-valued logic is not prevalent in the existing digital designs, SWSFET based circuits have good prospects for future multi-valued logical designs.

References

1. F. Jain, J. Chandy, and E. Heller, Proc. "Lester Eastman Conf. on High Performance Devices", Int. J. High Speed Electronics and Systems, Vol. 20, pp. 641-652, September 2011.
2. P. Gogna, M. Lingalugari, J. Chandy, E. Heller, E-S. Hasaneen and F. Jain, "Quaternary Logic and Applications Using Multiple Quantum Well Based SWSFETs", International Journal of VLSI design & Communication Systems (VLSICS) Vol. 3, No. 5, October 2012.
3. F. Jain 2015 JEM DOI number
4. R. Velampati, E-S. Hasaneen, E. Heller, and F. Jain, Floating Gate Nonvolatile Memory Using Individually Cladded Mono-dispersed Quantum Dots, IEEE Trans. VLSI, 25, pp. 1774-1781, May 2017.

5. Jain, Miller, Suarez, Chan, Karmakar, Al-Amoody, Chandy, Heller, "Spatial Wavefunction Switched (SWS) InGaAs FETs with II-VI Gate Insulators", Journal of Electronic Materials, Vol. 40, No. 8, pp. 1717-1726, 2011.

6. J. Rabey, A. Chandrakasaan and B. Nikolic, Design of Integrated Circuits, Prentice-Hall,2003.

7. B. Saman, P. Mirdha, M. Lingalugari, P. Gogna, and F. C. Jain., "Logic gates design and simulation using spatial wavefunction switched (SWS) FETs", International Journal for High-Speed Electronics and Systems, Vol. 24, Nos. 3 & 4 (2015), 1550008.

8. BSIM3v3 Manual, (Final Version), web site: rely.eecs.berkeley.edu or 128.32.156.10.

9. Analog Behavioral Modeling Applications reference manual. Cadence – application note, December 2009.

10. P. Gogna, M. Lingalugari, J. Chandy, F.C. Jain, E. Heller, and E. Hasaneen, Lester Eastman, Conference on High Performance Devices (LEC), IEEE Conference Publication (2012).

A Rapid Method Based on Fluorescence Spectroscopy for Meat Spoilage Detection

Binlin Wu[1,*], Kevin Dahlberg[1], Xin Gao[2], Jason Smith[1] and Jacob Bailin[1]

[1]*Physics Department and CSCU Center for Nanotechnology,*
Southern Connecticut State University, New Haven, CT 06515, USA
[2]*Natural Sciences Department, LaGuardia Community College, City University of New York,*
Long Island City, NY 11101, USA
*wub1@southernct.edu

Food spoilage is mainly caused by microorganisms, such as bacteria. In this study, we measure the autofluorescence in meat samples longitudinally over a week in an attempt to develop a method to rapidly detect meat spoilage using fluorescence spectroscopy. Meat food is a biological tissue, which contains intrinsic fluorophores, such as tryptophan, collagen, nicotinamide adenine dinucleotide (NADH) and flavin adenine dinucleotide (FAD) etc. As meat spoils, it undergoes various morphological and chemical changes. The concentrations of the native fluorophores present in a sample may change. In particular, the changes in NADH and FAD are associated with microbial metabolism, which is the most important process of the bacteria in food spoilage. Such changes may be revealed by fluorescence spectroscopy and used to indicate the status of meat spoilage. Therefore, such native fluorophores may be unique, reliable and non-subjective indicators for detection of spoiled meat. The results of the study show that the relative concentrations of all above fluorophores change as the meat samples kept in room temperature (~19°C) spoil. The changes become more rapidly after about two days. For the meat samples kept in a freezer (~-12°C), the changes are much less or even unnoticeable over a-week-long storage.

Keywords: Spoilage; bacteria; microbial metabolism; fluorescence spectroscopy; meat food; collagen; tryptophan; NADH.

1. Introduction

Food spoilage is an ever-increasing concern in the United States for supermarkets, restaurants and homes. Food spoilage is linked to several factors including temperature, humidity, oxygen, light and microorganisms [1]. In this study, grocery store meat, specifically chicken, is used. Meats are especially subject to the threat of bacterial microorganisms breaking down the tissue matter and causing food to spoil. If food is consumed after it has become spoiled by microorganisms, the consumer is at risk of becoming ill due to the byproducts from the degradation process. Several studies have found that microbial growth is the primary spoiling agent in meat samples [1, 2]. As a measure to protect businesses and families from the dangers associated with spoiling meat, this project aims to provide a rapid, user-friendly approach for detecting meat spoilage.

Many techniques exist for detecting spoilage in muscle foods such as chicken, pork, and beef. These techniques include, but are not limited to, Raman spectroscopy, magnetic resonance imaging, adenosine triphosphate (ATP) bioluminescence, and immunological or nucleic acid-based techniques [3]. These techniques prove to be time-consuming and require heavy involvement by a team of specialists. This makes it difficult to implement an economic and user-friendly mechanism for detecting meat spoilage. Fluorescence spectroscopy (FL) provides itself as a cost- and time-efficient technique for analyzing a collection of molecules in a complex mixture like that of meat food. In this study, we report on a mechanism involving native FL spectroscopy for the detection of meat spoilage.

In biological tissue, there are native (intrinsic) fluorophores such as collagen, elastin, nicotinamide adenine dinucleotide (NADH) and flavin adenine dinucleotide (FAD), and tryptophan [4]. Each fluorophore has characteristic excitation and emission spectra based on their chemical composition. A fluorescence spectrum of a sample acquired at a specific excitation wavelength is the mixture of the fluorescence spectra of all the intrinsic fluorophores with respect to that excitation wavelength. By probing meat samples with light at selective wavelengths, FL spectra of a meat sample can be measured to estimate the relative amount of fluorophores present either directly using the peak intensities or using an appropriate multivariate signal decomposition method such as nonnegative matrix factorization (NMF) [5-9]. During the process of spoilage, a tissue undergoes a variety of chemical changes, including the changes in the native fluorophores. These changes may be revealed using FL spectroscopy, which can then be used to indicate the status of spoilage in the tissue.

2. Experiments

Six chicken samples were prepared for this study. Three samples remained in air (~19°C) while the other three were kept in a standard freezer (~-12°C) and used for comparison in the results. The "air" samples remained in ambient conditions throughout the experiment with an expected higher spoilage rate. The samples stored in the freezer each had their own vessels for storage in the freezer, and were taken out only for measurements. The "freezer" samples were immediately returned to their vessels and stored in the freezer once data was collected. For the "air" samples, measurements were taken over 169 hours to show trends in spoilage as a function of fluorophore content in the sample. For the first four hours, data was collected every hour for the samples in air. Subsequent measurements were taken every two hours for the remaining four hours on day one. After one day, the samples were measured ever four hours for the remaining 145 hours of the study. The samples stored in the freezer were measured once every 24 hours over about the same time period (168 hours) as the samples stored at room temperature.

Perkin Elmer LS-55 Fluorescence Spectrometer with a data acquisition program FL WinLab™ was used in this study to acquire the fluorescence data. Optimal instrumental parameters were chosen at the beginning of the experiment to use the most dynamic range of the detector and fixed through the entire experiment. Five short wavelength cut-off filters (290-, 350-, 390-, 430-, and 515-nm) were used to isolate leaked excitation light in order to prevent tampering with emission results. During initial trials, we noticed a significant influence on the intensity of the emission peaks of collagen and NADH due to light leakage from the source. The filters were applied to mitigate this issue and allow for more robust results. We used an excitation wavelength of 260-nm with a 290-nm cut-off filter to probe the sample mainly for tryptophan, which has a peak around 340-nm. We then used an excitation wavelength of 320-nm with a 350-nm cut-off filter and an excitation wavelength of 340-nm with a 390-nm cut-off filter to mainly probe the collagen, NADH and FAD in the samples, which peak around 390-nm, 460-nm and 525-nm, respectively. The last two filters (430-, 515-nm) were used with excitation wavelengths 400-, 470-nm respectively. These filters should provide more refined peaks for NADH and FAD as the chosen wavelengths fall on selective excitation values for the two fluorophores.

We placed three samples in individual quartz cuvettes where they remained for the duration of the experiment. The goal was to reduce the work of tissue handling and the moving of the samples from their locations in order to take well-controlled measurements. The temperature in the room was measured throughout the day as data was being collected to give an average room temperature of ~19 degrees Celsius, and average freezer temperature of ~ -12 degrees Celsius. The emission spectra were acquired with a spectral range from 260-nm to 600-nm.

3. Analysis and Results

The excitation wavelengths used in the study are 260nm, 320nm, 340nm, 400nm, and 470nm, which were selected mainly based on the performance of the available cut-off filters. The emission peaks that are identified from the spectra using excitation wave-lengths, 260nm, 320nm, 400nm, and 470nm are considered to be mainly contributed by the fluorophores of tryptophan, collagen, NADH and FAD, respectively. All these wavelengths are listed in Table 1 below. The emission peak wavelengths of these fluorophores for "air" samples and "freezer" samples are slightly different.

Table 1. Characteristic fluorescence peak wavelengths with selective excitation wavelengths for different fluorophores.

Fluorophore	Excitation Wavelength (nm)	Emission peak wavelength (nm)	
		"air" samples	"freezer" samples
Tryptophan	260	336	336
Collagen	320	395	387
NADH	400	465	460
FAD	470	525	515

The fluorescence spectra of a representative "air" sample collected at t = 0 hour and t = 169 hour are shown in Fig. 1. The spectra are plotted both as excitation emission matrices (EEMs) and spectral profiles.

Similarly, the EEMs and the spectral profiles of a representative "freezer" sample collected at t = 0 hour and t = 168 hour are shown in Fig. 2.

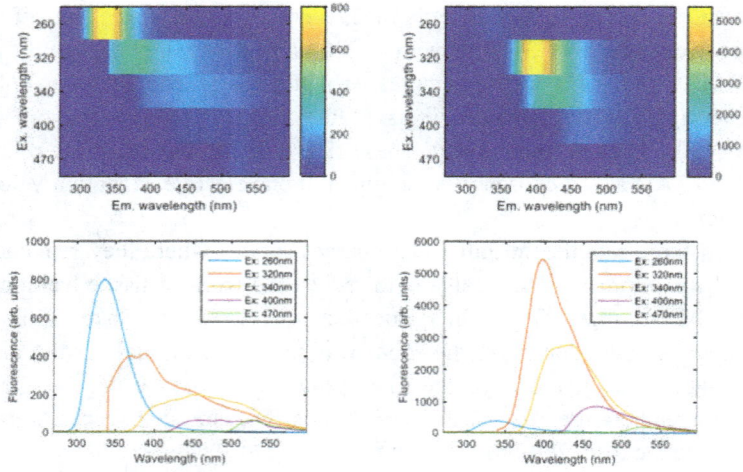

Fig. 1. (Top) Excitation emission matrix (EEM) for 0 hours (left) and 169 hours (right) from a representative sample stored in air. False color represents intensity in arbitrary units. (Bottom) Spectral profiles for generating EEM at 0 hours (left) and 169 hours (right).

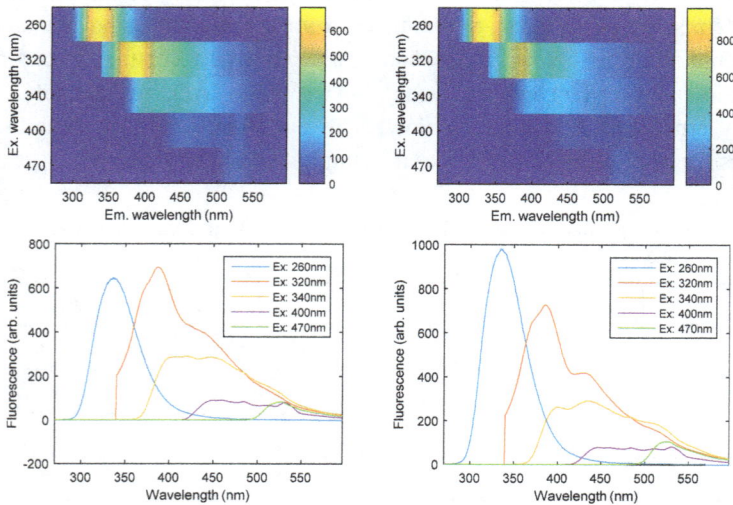

Fig. 2. (Top) EEM for 0 hours (left) and 168 hours (right) from all a representative sample stored in freezer. False color represents intensity in arbitrary units. (Bottom) Spectral profiles for generating corresponding EEMs above.

There are obvious changes in the spectra collected from the sample stored in air. But little changes or even unnoticeable changes are seen in the spectra for the sample stored in the freezer.

In this manuscript, we use the characteristic emission peak intensities of different fluorophores with respect to the selective excitation wavelength as shown above to estimate the relative concentrations of the fluorophores. Using the ratio between the characteristic peaks of the fluorophores, such as NADH and FAD, we estimate the relative changes in the concentrations of the fluorophores over the spoiling process. Graphs are created to display the change in the fluorescence signal over time for different fluorophores. The time evolution curves for tryptophan, collagen, NADH and FAD for the samples stored in air and freezer are shown in Figs. 3 and 4, respectively.

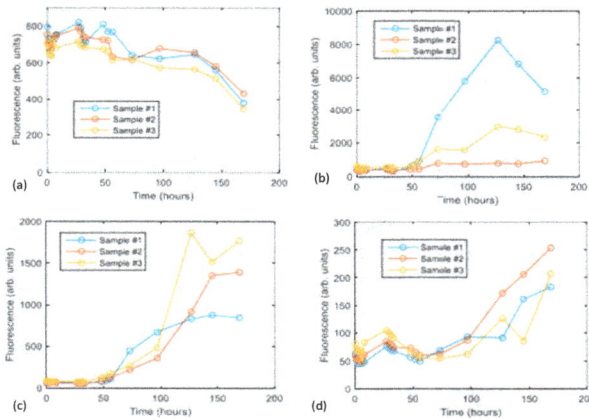

Fig. 3. Time evolution of fluorescence signal at 19 °C from (a) Tryptophan; (b) Collagen; (c) NADH; (d) FAD.

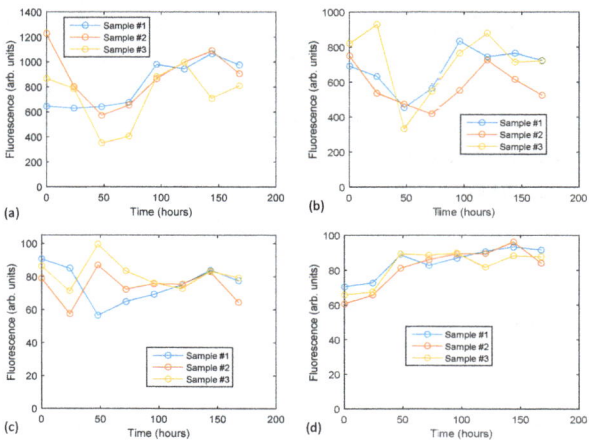

Fig. 4. Time evolution of fluorescence signal at -12 °C from (a) Tryptophan; (b) Collagen; (c) NADH; (d) FAD.

The same data sets were then used to create curves for the ratio of [NADH]/[FAD], i.e. redox ratio and [collagen]/[tryptophan]. The ratio curves for the samples stored in air and in freezer are shown in Fig. 5(a-b) and Fig. 5(c-d), respectively.

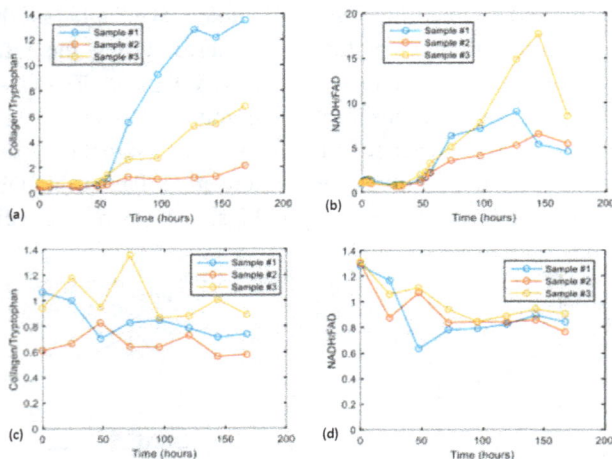

Fig. 5. (a) and (b) are ratio curves for three chicken samples held at 19 °C for the experiment. (c) and (d) are ratio curves for three chicken samples held at -12 °C for the experiment. (a) and (c): [collagen]/[tryptophan]; (b) and (d): [NADH]/[FAD].

4. Summary and Discussions

In this study, we measure the autofluorescence in meat samples longitudinally over a week in an attempt to develop a method to accurately and rapidly detect meat spoilage using fluorescence spectroscopy. In the experiments, the spectral non-uniformity of the excitation xenon lamp light source was pre-calibrated by Perkin-Elmer. Therefore, the system provided an effectively uniform illumination spectrum. Spectral non-uniformity also exists in the detector, i.e. photomultiplier tube (PMT), since the spectral response curve is generally wavelength dependent. This was not calibrated in the study either by the manufacturer or the authors. Using long-pass cut-off filters may also alter the spectral profiles since the transmission curve of the filters are generally wavelength dependent also. Even though the collected spectra can be distorted to some extent due to these reasons, it will not invalidate the method as long as all the measurements were taken under the same conditions.

The EEMs in Fig. 1 show a shift in the overall peak intensity down and to the right across the plot between the spectra taken at 0 hours to the spectra gathered at 169 hours of ambient exposure time. The false color intensity peak (bright yellow) shifts in this manner for a few reasons: (1) The tryptophan peak which is the global peak in the beginning is reduced over the time interval, suggesting that the abundance is decreasing in the sample. (2) The peak associated with 320-nm excitation is mainly due to increase in collagen, and becomes the global peak eventually. (3) The peak associated with

340-nm excitation starts off broad due to comparable signal level for collagen and NADH. Over time, the collagen signal increases more and dominates the peak. (4) The peak associated with 400-nm excitation is also broad over the range of emission wavelengths 430–540-nm. The main signal is considered to be from NADH and FAD at comparable levels. As the chicken spoils, the peak begins to squeeze to a narrow peak offset toward the 430-nm side of the initial peak likely due to the signal from NADH increasing relative to FAD. Both fluorophores are excited by the 400-nm incident light and thus the emission profile exhibits characteristics of both. A signal peak that favors NADH as the tissue samples spoils suggests that the relative abundance of NADH is increasing in the sample. (5) The curve generated by 470-nm excitation becomes dimmer over the course of the experiment due to the contrast change. The actual signal level increases as shown in Fig. 3(d). However, the increase may be due to influence by NADH, not by FAD. It needs further verification in future study. A decomposition method such as NMF may be used to separate signals due to NADH and FAD. In contrast to the "air" samples, Fig. 2 shows the spectra for the samples stored in freezer did not change significantly over the course of the experiment.

For the individual chemical signals, the first two days seem relatively stable as shown in Fig. 3. After the two-day period the signals changed significantly. This is different from what was reported in literature [10], where a dramatic change was observed at t = 4 hours for collagen and NADH. The data also show tryptophan decreases while the rest seem to increase. This is unusual for collagen as reported by others [10], where they reported a decrease in collagen relative abundance in spoiling pork tissue. It will be further investigated in future studies. For the samples stored in freezer, Fig. 4 shows unnoticeable change in the spectra, which is consistent with the freshness of the samples.

Due to systematic fluctuations, the ratios may be more robust characteristics of the spectra than peak intensities. An increase in the relative concentration ratio of [NADH]/[FAD] is shown in Fig. 5(a). This trend further demonstrates that the relative concentration of FAD in the samples decreases while the relative concentration of NADH increases as the meat undergoes the spoiling process, which is in agreement with the results in our previous study and literature [9, 10]. This change is considered to be mainly related to microbial metabolism, the most important process of the bacteria in food spoilage. Analysis of the [collagen]/[tryptophan] curve for the chicken samples in air also tends to increase during the spoilage process. This is a new discovery for these fluorophores, and suggests that tryptophan decreases in the sample as the bacteria breaks down the healthy tissue cells during spoilage. Tryptophan is an amino acid used in the biosynthesis of proteins. A typical byproduct of respiring bacteria left untreated, as in this study, is the formation of reactive oxygen species (ROS) [11]. In a process known as post-translational modification, the bacteria oxidize surrounding proteins in its new environment. This would result in a diminishing tryptophan abundance in the infected tissue. This result is also demonstrated in Fig. 3(a) for the time evolution of tryptophan signal in the spoiling meat.

The chicken samples that remain in the freezer show minimal change for all parameters over the seven-day period. This is expected since the cold environment of the freezer makes it more difficult for bacteria to penetrate and colonize the samples. However, the curve representing [NADH]/[FAD] shows a slight decrease which may be due to the chicken tissue physically degrading as a result of handling. Constantly removing the samples causes strain on the sample and can introduce a source of meat degradation that is not bacteria related. We can conclude from this comparison between the two data sets, both air and freezer, that the samples that remained in ambient conditions during the experiment were in fact undergoing the spoilage process. Further, we can confirm that the changes seen in Fig. 5(a) and 5(b) are indicative of meat spoilage in the chicken samples.

In our future study, we may use NMF to re-analyze the data to further investigate the inconsistencies displayed in Fig. 3 where we see collagen unexpectedly increasing [10], and the increase in the signal level where FAD peak is located. NMF may allow us to separate the fluorescence signal due to different fluorophores and remove background signal and noise effects that occur during the experiment, so that a better analysis may be provided. In the meantime, multiphoton microscopy (MPM) may be used to see the morphological changes in the samples, and further investigate the change in the amount of collagen, since MPM can also be a tool for second harmonic generation (SHG) imaging which can detect collagen very effectively [8, 12-17].

In summary, the project successfully detected changes in the relative concentrations of tryptophan, collagen, NADH, and FAD. Analysis of the ratios of the characteristic peak intensities in the collected spectra provides a rapid method for detecting spoilage of meat by detecting biochemical and morphological changes [18].

Acknowledgements

This study is supported in part by Faculty Creative Activity Research Grant at SCSU and CSU-AAUP Research Grant. Dr. B.W. also thanks MRRC grants at SCSU for providing support for attending the conference.

References

[1] Brooks, J. C., M. Alvarado, T. P. Stephens, J. D. Kellermeier, A. W. Tittor, M. F. Miller, and Brashears, M. M., "Spoilage and safety characteristics of ground beef packaged in traditional and modified atmosphere packages," J. Food Prot., 71(2), 293-301 (2008).

[2] Lambert, A. D., Smith, J. P., and Dodds, K. L., "Shelf life extension and microbiological safety of fresh meat — a review," Food Microbiol., 8(4), 267-297 (1991).

[3] Damez, J.-L. and Clerjon, S., "Meat quality assessment using biophysical methods related to meat structure," Meat Sci., 80, 132–149 (2008).

[4] Lakowicz, J. R., Principles of Fluorescence Spectroscopy, 3rd ed. New York: Springer US, 2006.

[5] Wu, B., Alrubaiee, M., Cai, W., Xu, M., and Gayen, S. K., "Diffuse optical Imaging using decomposition methods," Int. J. Opt., 2012, 185435 (2012).

[6] Wu, B., Gayen, S. K. and Xu, M., "Fluorescence spectroscopy using excitation and emission matrix for quantification of tissue native fluorophores and cancer diagnosis," Proc. SPIE, 8926, 89261M (2014).

[7] Wu, B. and S. K., "Fluorescence tomography of targets in a turbid medium using non-negative matrix factorization," Phys. Rev. E, 89, 042708 (2014).

[8] Wu, B., Li, G., Hao, M., and Mukherjee, S., "Non-invasive discrimination between pancreatic islets and exocrine cells using multiphoton microscopy," Proc. SPIE, 9329, 932935 (2015).

[9] Wu, B. and Dahlberg, K., "Measurement of food spoilage using fluorescence spectroscopy and imaging," Proc. SPIE, 9711, 97110X (2016).

[10] Pu, Y., Wang, W., and Alfano, R. R., "Optical Detection of Meat Spoilage Using Fluorescence Spectroscopy with Selective Excitation Wavelength," Appl. Spectrosc., 67(2), 210-213 (2013).

[11] Taylor, S. W., Fahy, E., Murray, J., Capaldi, R. A. and Ghosh, S. S., "Oxidative post-translational modification of tryptophan residues in cardiac mitochondrial proteins," J. Biol. Chem., 278(22), 19587-19590 (2003).

[12] Jain, M., Robinson, B. D., Wu, B. and Mukherjee, S., "Exploring multiphoton microscopy as a novel tool to differentiate chromophobe RCC from oncocytoma in fixed tissue sections," presented at the United States and Canadian Association of Pathology (USCAP) 2015 annual meeting, Boston MA, USA, 2015.

[13] Jain, M., Salvatore, S., Wu, B., Mukherjee, S. and Seshan, S. S., "Glomerular injury patterns as recognized by multiphoton microscopy in fixed tissue sections," presented at the United States and Canadian Association of Pathology (USCAP) 2015 annual meeting, Boston MA, USA, 2015.

[14] Jain, M., Salvatore, S., Robinson, B. D., Agarwaal, A., Wu, B., Scherr, D. S., Mukherjee, S. and Seshan, S. S., "Identifying signatures of normal and disease in freshly excised non-neoplastic kidney tissue with multiphoton microscopy," presented at the Renal Pathology in the XXXth Congress of the International Academy of Pathology (IAP), Bangkok, Thailand, 2014.

[15] Narula, N., Jain, M., Wu, B., Narula, J. and Mukherjee, S., "A component-by-component characterization of vulnerable plaques by multiphoton microscopy," presented at the United States and Canadian Association of Pathology (USCAP), San Diego, California, 2014.

[16] Wu, B., Mukherjee, S. and Jain, M., "A new method using multiphoton imaging and morphometric analysis for differentiating chromophobe renal cell carcinoma and oncocytoma kidney tumors," Proc. SPIE, 9712, 97121O (2016).

[17] Li, G., Wu, B., Ward, M. G., Chong, A. C. N., Mukherjee, S., Chen, S., and Hao, M., "Multifunctional in vivo imaging of pancreatic islets during diabetes development," J. Cell Sci., 129(14), 2865-2875 (2016).

[18] Wu, B., Dahlberg, K., Gao, X., Smith, J. and Bailin, J., "Rapid measurement of meat spoilage using fluorescence spectroscopy," Proc. SPIE 10068, Imaging, Manipulation, and Analysis of Biomolecules, Cells, and Tissues XV, 1006820 (2017).

Optical Biopsy for Prostate Cancer Diagnosis Using Fluorescence Spectroscopy

Binlin Wu[1,*], Xin Gao[2] and Jason Smith[1]

[1]*Physics Department and CSCU Center for Nanotechnology,*
Southern Connecticut State University, New Haven, CT 06515, USA
[2]*Natural Sciences Department, LaGuardia Community College, City University of New York,*
Long Island City, NY 11101, USA
[*]wub1@southernct.edu

Native fluorescence spectra are acquired from fresh normal and cancerous human prostate tissues. The fluorescence data are analyzed using an unsupervised machine learning algorithm such as non-negative matrix factorization. The nonnegative spectral components are retrieved and attributed to the native fluorophores such as collagen, reduced nicotinamide adenine dinucleotide (NADH), and flavin adenine dinucleotide (FAD) in tissue. The retrieved scores of the components are used to estimate the relative concentrations of the native fluorophores such as NADH and FAD and the redox ratio. A supervised machine learning algorithm such as support vector machine (SVM) is used to classify normal and cancerous tissue samples based on either the relative concentrations of NADH and FAD or the redox ratio alone. Various statistical measures such as sensitivity, specificity, and accuracy, along with the area under receiver operating characteristic (ROC) curve are used to show the classification performance. A cross validation method such as leave-one-out is used to further evaluate the predictive performance of the SVM classifier to avoid bias due to overfitting, and the accuracy was found to be 93.3%.

Keywords: Prostate cancer; fluorescence spectroscopy; intrinsic tissue emission; optical biopsy; redox ratio; nonnegative matrix factorization; support vector machine; machine learning; NADH; FAD.

1. Introduction

Aside from skin cancer, prostate cancer is the most common cancer in American men [1]. About 164,690 new cases of prostate cancer will be diagnosed, and about 29,430 men will die of prostate cancer in the United States in 2018 according to American Cancer Society's estimates.

The traditional method to diagnose prostate cancer is biopsy, which is considered the gold standard. A biopsy is a process to remove a small piece of tissue sample from the body of a patient, thereby it can be stained with haematoxylin and eosin (H&E) or by other special staining means and then analyzed by a pathologist using a microscope. Such a process is invasive, time consuming, and subjective due to the judgement of the pathologist. Optical biopsy (OB) is an alternative method and has attracted extensive attentions in the past decades. OB is a collection of diagnostic techniques that diagnose a tissue sample by measuring its optical properties using optical methods such as fluorescence spectroscopy [2, 3], Raman spectroscopy [4-6], and multiphoton microscopy [7-14]. OB is non-invasive and rapid, and quantifies tissue properties for diagnosis.

Of various OB methods, fluorescence spectroscopy is a technique that can perform cancer diagnosis by measuring the autofluorescence spectra, i.e. tissue intrinsic emission generated by the native fluorophores in tissue. For a cancer that develops in prostate tissue, with its grade advances [15], the tissue undergoes a series of biochemical and morphological changes [15, 16]. Reduced nicotinamide adenine dinucleotide (NADH), and flavin adenine dinucleotide (FAD) are two typical fluorophores that are involved in the oxidation of fuel molecules in cellular metabolism. The concentrations of these metabolites change as cancer develops [2, 17, 18]. The ratio between NADH and FAD in various forms, is called oxidation-reduction ratio or redox ratio, and related to the metabolism level [3, 19-22], and therefore has been used to detect the malignancy of tissue [3, 17, 18, 23-25]. Here we simply use NADH to represent signals from reduced form of both nicotinamide adenine dinucleotide (NAD$^+$) and nicotinamide adenine dinucleotide phosphate (NADP$^+$), which together is denoted by NAD(P)H, since their emission spectra are identical [26]. In fact, the cellular content and fluorescence quantum yield of NADH are typically higher than that of NADPH, therefore we expect that our results mainly indicated the concentration of NADH [27].

In this study, we evaluate the efficacy of prostate cancer diagnosis using relative concentrations of NADH and FAD as well as the redox ratio revealed from autofluorescence spectra of fresh frozen prostate tissue specimens.

2. Experimental and Analytical Methods

Fifteen fresh prostate tissue samples including 8 normal and 7 cancerous samples were purchased from tissue bank Cooperative Human Tissue Network (CHTN) and stored in -80°C freezer before experiments. Fluorescence spectra were acquired using a Perkin Elmer LS-50 fluorescence spectrometer. The wavelength used for excitation was 340-nm. The spectral range for detection was 360-nm to 650-nm with 1nm spectral resolution. Proper instrumental parameters were chosen at the beginning of the experiment to optimize the signal. The spectral data were then used for subsequent analysis.

The spectral data were analyzed using a decomposition method non-negative matrix factorization (NMF) [2, 28-31]. NMF is an unsupervised machine learning algorithm that is to separate component signals and their abundances from mixed signals. In matrix notation, it may be expressed as follows. The mixed signals in a matrix $X = WH+N$, where columns of W are the components or retrieved emission spectra of native fluorophores in this study; H is a mixing matrix and its rows are the scores for the components and may be considered as the relative concentrations of the fluorophores; columns of X are the mixed signals and correspond to tissue sample spectra in this study; and N is the noise. The objective of NMF is to factorize the matrix X into W and H, where W and H are non-negative [32]. NMF learns parts-based representation of the signal and find the hidden parts which may be more recognizable [28]. Since the light signal and the concentrations of the molecules are naturally nonnegative values, NMF may be able to retrieve the components corresponding to individual fluorophores [28, 33].

NMF has been extensively used in diverse applications such as, facial image recognition [28], genetic and molecular pattern discovery [34], spectral data analysis [35], cancer class discovery [36] and diffuse optical imaging [29].

Once the component spectra along with the relative concentrations of the intrinsic fluorophores NADH and FAD were retrieved, the redox ratio defined as [FAD]/([NADH]+[FAD]) was calculated and used to characterize the tissue samples. A supervised machine learning algorithm support vector machine (SVM) with a linear kernel [2, 11, 12, 37-39] was used to classify normal and cancerous tissues. The features used for classification were either relative concentrations of NADH and FAD or the redox ratio alone. Statistical measures including sensitivity, specificity, and accuracy, as well as area under receiver operating characteristic (ROC) curve [12] i.e. AUROC were used to evaluate the performance of the classification. ROC curve is a plot of true positive rate (TPR) vs. false positive rate (FPR), i.e. sensitivity vs. 1-specificity. ROC is generated by varying a decision threshold to separate the positive and negative classes, such as a threshold for posterior probabilities calculated from SVM scores [40]. To avoid bias due to overfitting, a cross validation method such as leave-one-out cross validation (LOOCV) [11, 12, 39] was used and compared with re-substitution validation method.

3. Analysis and Results

Typical spectra of normal and cancerous prostate tissue samples are shown below in Fig. 1. A profile difference around 460-nm can be seen in the spectra, which corresponds to NADH.

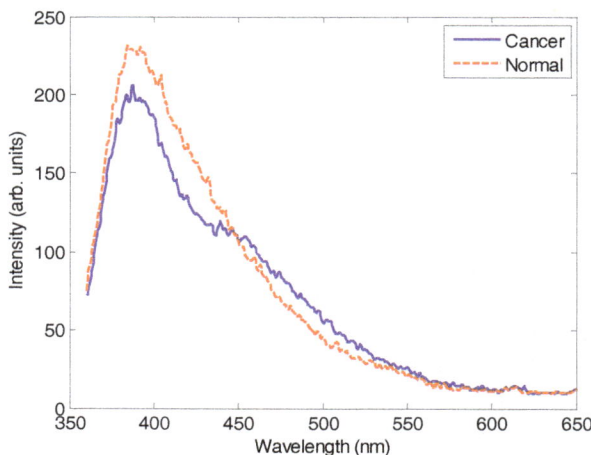

Fig. 1. Typical spectra of normal and cancerous prostate tissue.

The NMF retrieved component spectra are shown in Fig. 2. The raw tissue fluorescence spectra were decomposed into six components. The retrieved signals for NADH and FAD are quite low compared to collagen, since 340-nm is not the optimal excitation wavelength for these two fluorophores, but near the optimal excitation wavelength for collagen [3]. This is clearly seen in the raw tissue spectra, which show the most prominent peaks around 380-nm. The component spectra that best correlate with NADH and FAD were identified using the peak wavelengths, i.e. ~460-nm and ~525-nm, and labeled in Fig. 2.

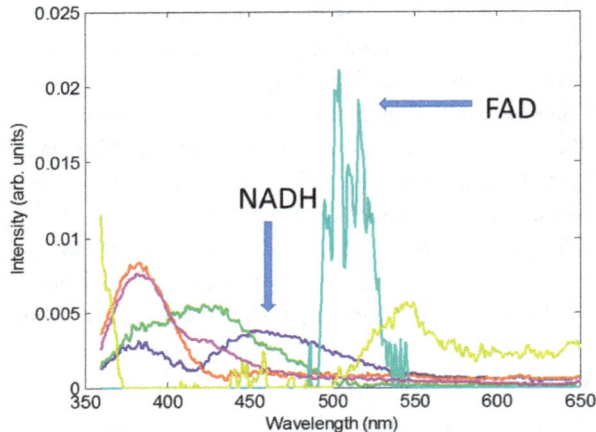

Fig. 2. NMF component spectra retrieved from the fifteen raw spectra. The component spectra for NADH and FAD are labeled.

The component scores corresponding to NADH and FAD retrieved by NMF were considered to be the relative concentrations of the fluorophores, and used for classi-fication using a linear SVM. A SVM classifier trained with all the 15 samples is shown in Fig. 3(a). If re-substitution validation [11, 12, 41] is used, i.e. the classifier is applied on the same 15 samples, sensitivity, specificity and accuracy were all found to be 100%. In other words, a separation line was found for this dataset to perfectly separate normal and cancerous tissue samples. A ROC curve was generated and shown in Fig. 3(b). The AUROC was also calculated to be 100%. If LOOCV was applied, the sensitivity, specificity, accuracy reduced to 85.7%, 100%, and 93.3%.

Based on NMF-retrieved component scores of NADH and FAD, the redox ratio was then calculated for all tissue samples. SVM classification based on the redox ratio was also performed. The result is shown in Fig. 4(a). The corresponding ROC curve was also generated and shown in Fig. 4(b). Sensitivity, specificity and accuracy were calculated to be 85.7%, 100%, and 93.3%, with AUROC 89.3%, using re-substitution validation method. With LOOCV, the sensitivity, specificity and accuracy were found to be the same as those using re-substitution method.

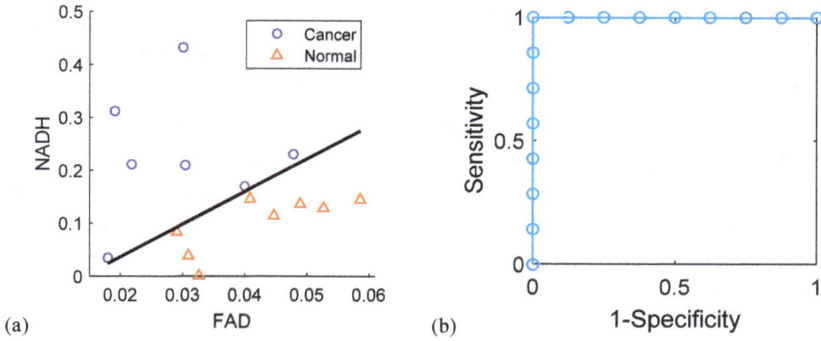

Fig. 3. (a) SVM classification based on NMF-retrieved component weights for NADH and FAD. (b) ROC curve.

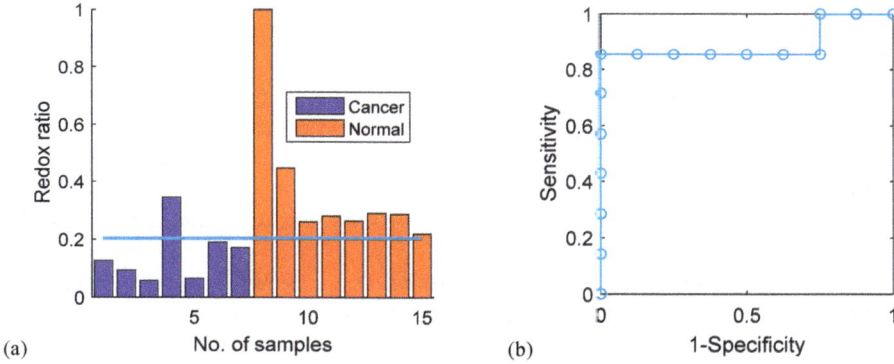

Fig. 4. (a) SVM classification based on redox ratio. (b) ROC curve.

4. Summary and Discussions

In summary, fluorescence spectroscopy was used to measure the intrinsic emission from normal and cancerous prostate tissue samples [42, 43]. Unsupervised machine learning algorithm NMF was used to decompose the intrinsic tissue fluorescence spectra. The component spectra were considered to be the emission spectra of the native fluorophores in prostate tissue samples, and the component scores were estimates of the relative concentrations. The relative concentrations of NADH and FAD as well as the redox ratio calculated from these relative concentrations were used to as the features to distinguish normal and cancerous tissue samples. A linear SVM was used for classification. The performance of the separation was evaluated using statistical measures including sensitivity, specificity, accuracy, and AUROC. LOOCV was used to avoid bias in classification due to overfitting. Using LOOCV, a classification accuracy 93.3% was achieved. The results showed using either the relative concentrations of NADH and FAD, or the redox ratio can distinguish normal and cancerous prostate tissues with high accuracy.

Even though the separation based on relative concentrations of NADH and FAD showed higher accuracy than that based on redox ratio when re-substitution validation was used, this could be due to overfitting. Further analysis showed using NADH vs FAD provided same accuracy as using redox ratio when LOOCV was used. A larger-size sample need to be used in future study to verify these results.

Since the retrieved component spectrum of NADH has some collagen signal mixed in it, the result may be partially affected by the collagen in tissue. Further experiments using other wavelengths may be performed to better separate the component spectra for different fluorophores, or using excitation emission matrix [2, 3].

Acknowledgements

This study is supported in part by Faculty Creative Activity Research Grant at SCSU and CSU-AAUP Research Grant.

References

[1] http://www.cancer.org/cancer/prostatecancer/detailedguide/prostate-cancer-key-statistics.

[2] Wu, B., Gayen, S. K., and Xu, M., "Fluorescence spectroscopy using excitation and emission matrix for quantification of tissue native fluorophores and cancer diagnosis," *Proc. SPIE* **8926**, 89261M (2014).

[3] Lakowicz, J. R., *Principles of Fluorescence Spectroscopy*, 3rd ed., Springer US, New York (2006).

[4] Liu, C.-H., Sha Glasman, W. L., Zhu, H. R., Akins, D. L., Deckelbaum, L. I., Stetz, M. L., O'Brien, K., Scott, J., and Alfano, R. R., "Near-IR Fourier transform Raman spectroscopy of normal and atherosclerotic human aorta," *Laser in the Life Sciences* **43**, 257–264 (1992).

[5] Liu, C.-H., Zhou, Y., Sun, Y., Li, J. Y., Zhou, L. X., Boydston-White, S., Masilamani, V., Zhu, K., Pu, K. Y., and Alfano, R. R., "Resonance Raman and Raman spectroscopy for breast cancer detection," *Technol. Cancer Res. Treat.* **12**, 371–382 (2013).

[6] Liu, C.-H., Sriramoju, V., Boydston-White, S., Wu, B., Zhang, C., Pei, Z., Sordillo, L., Beckman, H., and Alfano, R. R., "Resonance Raman of BCC and normal skin," *Proc. SPIE* **10060**, 100601B (2017).

[7] Jain, M., Robinson, B. D., Wu, B., and Mukherjee, S., "Exploring multiphoton microscopy as a novel tool to differentiate chromophobe RCC from oncocytoma in fixed tissue sections," in *United States and Canadian Association of Pathology (USCAP) 2015 annual meeting*, Paper 920, Boston MA, USA (2015).

[8] Jain, M., Salvatore, S., Wu, B., Mukherjee, S., and Seshan, S. S., "Glomerular injury patterns as recognized by multiphoton microscopy in fixed tissue sections," in *United States and Canadian Association of Pathology (USCAP) 2015 annual meeting*, Paper 1625, Boston MA, USA (2015).

[9] Jain, M., Salvatore, S., Robinson, B. D., Agarwaal, A., Wu, B., Scherr, D. S., Mukherjee, S., and Seshan, S. S., "Identifying signatures of normal and disease in freshly excised non-neoplastic kidney tissue with multiphoton microscopy," in *Renal Pathology in the XXXth Congress of the International Academy of Pathology (IAP)*, pp. S131-S132, Bangkok, Thailand (2014).

[10] Narula, N., Jain, M., Wu, B., Narula, J., and Mukherjee, S., "A component-by-component characterization of vulnerable plaques by multiphoton microscopy," in *United States and Canadian Association of Pathology (USCAP)*, San Diego, California (2014).

[11] Wu, B., Nebylitsa, S. V., Mukherjee, S., and Jain, M., "Quantitative diagnosis of bladder cancer by morphometric analysis of HE images," *Proc. SPIE* **9303**, 930317 (2015).

[12] Wu, B., Mukherjee, S., and Jain, M., "A new method using multiphoton imaging and morphometric analysis for differentiating chromophobe renal cell carcinoma and oncocytoma kidney tumors," *Proc. SPIE* **9712**, 97121O (2016).

[13] Jain, M., Robinson, B. D., Wu, B., Khani, F., and Mukherjee, S., "Exploring multiphoton microscopy as a novel tool to differentiate chromophobe renal cell carcinoma from oncocytoma in fixed tissue sections," *Arch. Pathol. Lab. Med.* **142**, 383–390 (2018).

[14] Jain, M., Wu, B., Pisapia, D., Salvatore, S., Mukherjee, S., and Narula, N., "A component-by-component characterisation of high-risk atherosclerotic plaques by multiphoton microscopic imaging," *J. Microsc.* **268**, 39–44 (2017).

[15] Gleason, D. F., and Mellinger, G. T., "Prediction of prognosis for prostatic adenocarcinoma by combined histological grading and clinical staging," *J. Urol. (Baltimore)* **111**, 58–64 (1974).

[16] Morrison, C., Thornhill, J., and Gaffney, E., "The connective tissue framework in the normal prostate, BPH and prostate cancer: analysis by scanning electron microscopy after cellular digestion," *Urol. Res.* **28**, 304–307 (2000).

[17] Skala, M. C., Riching, K. M., Gendron-Fitzpatrick, A., Eickhoff, J., Eliceiri, K. W., White, J. G., and Ramanujam, N., "*In vivo* multiphoton microscopy of NADH and FAD redox states, fluorescence lifetimes, and cellular morphology in precancerous epithelia," *PNAS* **104**, 19494–19499 (2007).

[18] Skala, M., and Ramanujam, N., "Multiphoton Redox Ratio Imaging for Metabolic Monitoring *In Vivo*," in *Advanced Protocols in Oxidative Stress II*, D. Armstrong, Ed., pp. 155–162, Humana Press (2010).

[19] Balaban, R. S., and Mandel, L. J., "Optical methods for the study of metabolism in intact cells," in *Noninvasive Techniques in Cell Biology*, J. K. Foskett, and S. Grinstein, Eds., Wiley-Liss, New York (1990).

[20] Chance, B., "Optical Method," *Annu. Rev. Biophys. Biophys. Chem.* **20**, 1–28 (1991).

[21] Masters, B. R., "Noninvasive corneal redox fluorometry," *Curr. Top. Eye Res.* **4**, 139–200 (1984).

[22] Masters, B. R., and Chance, B., "Redox confocal imaging: intrinsic fluorescent probes of cellular metabolism," in *Fluorescent and Luminescent Probes for Biological Activity*, W. T. Mason, Ed., pp. 44–57, Academic Press, New York (1993).

[23] Chance, B., Schoener, B., Oshino, R., Itshak, F., and Nakase, Y., "Oxidation-reduction ratio studies of mitochondria in freeze-trapped samples. NADH and flavoprotein fluorescence signals," *J. Biol. Chem.* **254**, 4764–4771 (1979).

[24] Chance, B., "Metabolic heterogeneities in rapidly metabolizing tissues," *J. Appl. Cardiol.* **4**, 207–221 (1989).

[25] Heintzelman, D. L., Lotan, R., and Richards-Kortum, R. R., "Characterization of the Autofluorescence of Polymorphonuclear Leukocytes, Mononuclear Leukocytes and Cervical Epithelial Cancer Cells for Improved Spectroscopic Discrimination of Inflammation from Dysplasia," *Photochem. and Photobiol.* **71**, 327–332 (2000).

[26] Huang, S., Heikal, A. A., and Webb, W. W., "Two-Photon Fluorescence Spectroscopy and Microscopy of NAD(P)H and Flavoprotein," *Biophysical Journal* **82**, 2811–2825 (2002).

[27] Mujat, C., Kim, Y. L., Greiner, C., Backman, V., Baldwin, A., Feld, M., Levitt, J. M., Munger, K., Georgakoudi, I., Tian, F., Stucenski, L. A., and Hunter, M., "Endogenous optical biomarkers of normal and human papillomavirus immortalized epithelial cells," *Int. J. Cancer* **122**, 363–371 (2008).

[28] Lee, D. D., and Seung, H. S., "Learning the parts of objects by non-negative matrix factorization," *Nature* **401**, 788–791 (1999).

[29] Wu, B., Alrubaiee, M., Cai, W., Xu, M., and Gayen, S. K., "Diffuse optical Imaging using decomposition methods," *Int. J. Opt.* **2012**, 185435 (2012).

[30] Wu, B., S. K., "Fluorescence tomography of targets in a turbid medium using non-negative matrix factorization," *Phys. Rev. E* **89**, 042708 (2014).

[31] Wu, B., and Dahlberg, K., "Measurement of food spoilage using fluorescence spectroscopy and imaging," *Proc. SPIE* **9711**, 97110X (2016).

[32] Berry, M. W., Browne, M., Langville, A. N., Pauca, V. P., and Plemmons, R. J., "Algorithms and applications for approximate nonnegative matrix factorization," *Comp. Stat. Data Anal.* **52**, 155–173 (2007).

[33] Georgakoudi, I., Jacobson, B. C., Muller, M. G., Sheets, E. E., Badizadegan, K., Carr-Locke, D. L., Crum, C. P., Boone, C. W., Dasari, R. R., Dam, J. V., and Feld, M. S., "NAD(P)H and Collagen as *in vivo* Quantitative Fluorescent Biomarkers of Epithelial Precancerous Changes," *Cancer Res.* **62**, 682–687 (2002).

[34] Brunet, J.-P., Tamayo, P., Golub, T. R., and Mesirov, J. P., "Metagenes and molecular pattern discovery using matrix factorization," *PNAS* **101**, 4164–4169 (2004).

[35] Pauca, V. P., Piper, J., and Plemmons, R. J., "Nonnegative matrix factorization for spectral data analysis," *Lin. Alg. Appl.* **416**, 29–47 (2006).

[36] Gao, Y., and Church, G., "Improving molecular cancer class discovery through sparse non-negative matrix factorization," *Bioinformatics* **21**, 3970–3975 (2005).

[37] Cortes, C., and Vapnik, V., "Support-vector networks," *Machine Learning* **20**, 273 (1995).

[38] Press, W. H., Teukolsky, S. A., Vetterling, W. T., and Flannery, B. P., "Support vector machines," in *Numerical Recipes: The Art of Scientific Computing*, Cambridge University Press New York, NY (2007).

[39] Wu, B., Li, G., Hao, M., and Mukherjee, S., "Non-invasive discrimination between pancreatic islets and exocrine cells using multiphoton microscopy," *Proc. SPIE* **9329**, 932935 (2015).

[40] Platt, J. C., "Probabilistic outputs for support vector machines and comparisons to regularized likelihood methods," in *Advances in Large Margin Classifiers*, A. J. Smola, P. Bartlett, B. Schölkopf, and D. Schuurmans, Eds., pp. 61–74, MIT Press, Cambridge, MA (1999).

[41] Refaeilzadeh, P., Tang, L., and Liu, H., "Cross Validation," in *Encyclopedia of Database Systems*, M. T. Özsu, and L. Liu, Eds., pp. 532–538, Springer (2009).

[42] Wu, B., Gao, X., Smith, J., and Bailin, J., "Optical biopsy using fluorescence spectroscopy for prostate cancer diagnosis," *Proc. SPIE 10038, Therapeutics and Diagnostics in Urology: Lasers, Robotics, Minimally Invasive, and Advanced Biomedical Devices*, 100380U (2017).

[43] Xue, J., Pu, Y., Smith, J., Gao, X., and Wu, B., "Machine learning based analysis of human prostate cancer cell lines at different metastatic ability using native fluorescence spectroscopy with selective excitation wavelength," *Proc. SPIE 10504, Biophysics, Biology and Biophotonics III: the Crossroads*, 105040L (2018).

Fast and Reversible Chemiresistive Sensors for Robust Detection of Organic Vapors Using Oleylamine-Functionalized Palladium Nanoparticles

Tuo Gao[1], Yongchen Wang[2], Yi Luo[2], Chengwu Zhang[1], Zachariah Pittman[1], Alexandra Oliveira[1], Howard Craig[1], Jing Zhao[2,3] and Brian G. Willis[1,3,*]

[1]Department of Chemical and Biomolecular Engineering, [2]Department of Chemistry,
[3]Institute of Materials Science, University of Connecticut, Storrs, Connecticut 06269, USA
*brian.willis@uconn.edu

Chemiresistive sensors fabricated by oleylamine-functionalized palladium nanoparticles (OLA-PdNP) have been studied in hydrogen sensing, but not so much in organic vapor sensing. Like the extensively studied gold nanoparticles-based gas sensors, palladium nanoparticles also give the ease of surface modification and large surface-area-to-volume ratio. In this study, we demonstrate an OLA-PdNP chemiresistor array with robust sensor responses (1-15% of $\Delta R/R_0$) and accurate discrimination of six organic vapors at a concentration of $p/p_0 = 0.2$, using principal component analysis (PCA). Each microfabricated 36 mm^2 chip has 36 individual sensors. By incorporating multiple sensors on one chip, the sensor response gives a distinguishable pattern for each analyte. From this study, an electronic chemical spectrometer can be further developed by incorporating many types of ligands on palladium metal core to enhance sensor accuracy and precision.

Keywords: Chemiresistors; palladium nanoparticles; nanofabrication; principal component analysis.

1. Introduction

High-speed, miniaturized, portable chemical sensors have a wide range of application in environmental monitoring, homeland security, food safety, and health screening. Specifically, metal nanoparticles with self-assembled organo-functional monolayers (SAM) have provided the ability to detect gases and organic vapors with fast responses. As rapid-developing sensing elements, organo-functionalized metal nanoparticles (NP) transduce vapor sorption in SAM layers into measurable electrical signals. Compared with other traditional sensing elements, such as metal oxides and polymers, metal nanoparticles offer several advantages including smaller sizes, larger surface area to volume ratio, straightforward synthetic methods, and versatility in ligand functionality [1, 2]. To achieve good selectivity and sensitivity toward various analytes, diverse metal-ligand pairs are essential to construct high-order chemiresistive sensor arrays.

Chemical sensors using organo-functionalized palladium nanoparticles (PdNP) have been studied on hydrogen sensing over the past decades. Moreno *et al.* synthesized palladium nanoparticles using octylamine and hexanethiol as ligands and their findings indicated that improved stability and reactivity in hydrogen can be achieved by using mixed clusters with both types of ligands [3]. In spite of many successful advances in hydrogen sensing, there are fewer reports of organic chemical vapors sensing using PdNP. Most of the organic vapor chemiresistive sensors focused on functionalized gold nanoparticles, platinum nanoparticles, as well as alloys [1, 2] Previous studies in our group had demonstrated chemiresistive sensors fabricated by polynucleotide-functionalized gold nanoparticles using five single-strand DNA sequences [4].

In this study, we present the integration of oleylamine-functionalized palladium nanoparticles (OLA-PdNP) into a 36-device array of chemiresistive sensors to detect six common laboratory organic vapors. Controlled assemblies of palladium nanoparticles were achieved by drop-casting and dielectrophoresis (DEP) into 2-μm and 50-nm gaps fabricated using photolithography and electron-beam lithography, respectively. Simultaneous measurement of real time sensitivity, expressed as $\Delta R/R_0$, were performed for eight different devices integrated on the same chip. In DEP experiments, the trapping density of nanoparticles around the sensing region can be tuned by changing applied frequency, peak-to-peak voltage, and processing time that will lead to a variation in device baseline resistances. Working sensors with 1-15% $\Delta R/R_0$ were achieved with a baseline resistance between 1-100 Ω. Sensing performance is evaluated by principal component analysis (PCA) that illustrates unique fingerprints for each analyte. In addition, sensitivity response trends scale with baseline resistance. These results provide insights into selection of the most ideal metal-ligands to sense analytes in open air environments.

2. Experimental Section

This section describes the synthetic method of amine-functionalized palladium nanoparticles, fabrication of chemiresistor arrays, and the vapor sensing experiment. An illustration of the experimental set up is shown in Fig. 1.

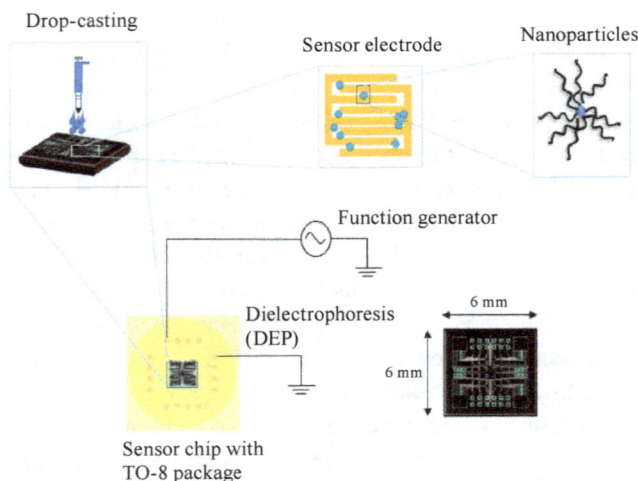

Fig. 1. Schematic diagram of nanoparticle assembly on chemiresistive devices.

2.1. *Synthesis of Oleylamine-Functionalized Palladium Nanoparticle*

In a typical synthesis, Pd(acac)$_2$ (0.3 mmol, 91.2 mg), oleylamine (20 mmol, 5 mL) and a magnetic stir bar were added to a 25 mL three-neck flask under nitrogen protection. The reaction mixture was heated to 100 °C and kept for 20 min. After cooling down to 80 °C, the product was purified by adding ethanol and centrifuging at 2000 rpm for 2 min. Then the particles were kept in ethanol for further use. The average concentration of the as-synthesized PdNP solution was 3 mg/mL with an average individual particle size of 4 nm. Clusters of PdNPs with an average size of 200 nm were formed and stabilized over time. Figure 2 shows TEM image of oleylamine-functionalized PdNPs and SEM image of same type of nanoparticles drop-cast onto microelectrodes.

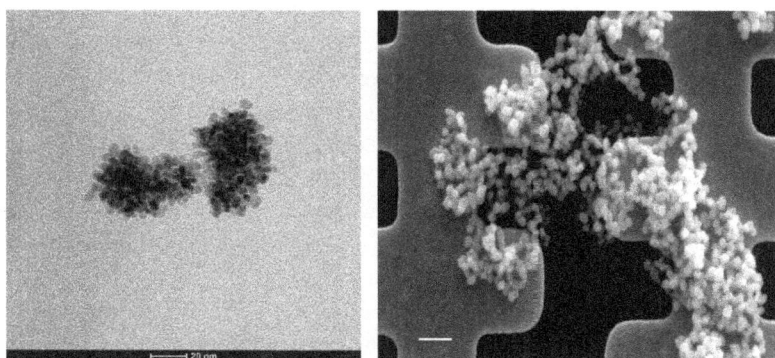

Fig. 2. Left: TEM of OLA-PdNP; Right: SEM of OLA-PdNP between electrodes (scale bar = 1 μm).

2.2. *Fabrication of Chemiresistive Devices*

Two types of chemiresistive sensors, 50 nm gap castellated nanodevices and 2 μm gap castellated microdevices were fabricated using lithography processes on silicon wafer substrates with 300 nm thermal oxide. Electron-beam lithography was used to define nanoscale features. The active sensing regions for each device occupied approximately 30 μm^2. For nanodevices, each electrode had 50 rectangular tip pairs. A total number of 36 individual addressable devices were defined within a 2 × 2 mm^2 area. A thin-film of 40 nm gold with 5 nm titanium adhesion layer were deposited onto the wafer using electron-beam evaporation followed by liftoff processes. Optical photolithography was used to define metal contacts to 50 nm gap castellated nanodevices and also to fabricate 2 μm castellated microdevices. Metal electrodes were defined by liftoff after the deposition of 200 nm gold with 10 nm titanium to obtain a final sensor chip. Compared with earlier generations of sensor arrays, we use a more space-efficient design. Instead of creating contact pads for each device, a common ground at each corner was shared by 9 devices.

2.3. *Vapor Sensing Experiments*

Nanoparticles were fabricated into the electrode gaps by two approaches: direct drop-casting and dielectrophoresis (DEP). In the former method, 2 μL of nanoparticle solutions were directly drop-cast onto the center of a chip using micropipettes. The droplets spread out to wet the entire chip and cover all devices. Once the solvent evaporated, the entire chip was rinsed with ethanol and subsequently dried by flowing dry nitrogen to remove stray particles. In the latter method, an AC voltage was applied by a function generator (Model 33220A, Agilent) on each device to provide an electric driving force to attract nanoparticles from the bulk solution. The voltage was turned on after micro-pipetting 2 µL nanoparticle solutions for a controlled time and frequency. The remainder solution was blown off the chip by dry nitrogen. Prior to sensing experiments, the chips were placed in the dark for 1 hour for the nanoparticles to stabilize.

For 36-device sensors, the chip was mounted on a TO-8 package (Spectrum Semiconductor Materials) whose leads were wire bonded (Model 747630E, WestBond) to the contact pads. The TO-8 package was inserted into a solderless breadboard (3M) to create electrical connection to the switch matrix (Model 7002, Keithley) and source meter (Model 2612, Keithley). Saturated organic vapors were delivered to the sensor at a known rate through a syringe pump (Model 100, Cole-Parmer). Simultaneous sensor responses from 8 devices were recorded upon pulsing of each vapor analyte. The devices were purged with dry nitrogen between experiments.

Sensing data were obtained using a custom-written LabVIEW program. The assemblies of nanoparticles on devices were imaged using a scanning electron microscope (Teneo, FEI). To evaluate the differentiation of organic vapors, principal component analysis (PCA) was performed. The baseline resistance for each sensor was normalized with a Savitzky-Golay filtering method.

3. Results and Discussion

In this section, we investigate the trapping of nanoparticle and evaluate sensor performance with OLA-PdNP. The average $\Delta R/R$ are tabulated into the PCA matrix to assess the discrimination of six common organic vapors.

3.1. *Dielectrophoresis (DEP)-Assisted Nanoparticle Assembly*

Castellated nanostructures with sharp rectangular corners were fabricated as the electrodes because of the ease of trapping nanoparticles, as the non-uniform electric field is enhanced at edges and corners [5]. Consequently, polarizable nanoparticles are assembled between electrodes by the dielectrophoretic force. Figure 3 shows the dielectrophoresis results for nanodevices using gold nanorods coated by hexadecyltrimethylammonium bromide (CTAB) double layer.

DEP experiments were performed using a sinusoidal waveform at a fixed AC frequency of 1 MHz and peak-to-peak voltage of 4 V. The processing time was varied at 1, 2, and 4 min. At 1 min deposition, the nanoparticles are mostly isolated and trapped near the edge of nanoelectrodes. At 2 min deposition, the number of captured nanoparticles increases, and clusters are formed between electrode tips. The resistance dropped to 7 kΩ. At 4 min deposition, nanoparticles are filled in electrode gaps and the outer edges. Interestingly, the

Fig. 3. SEM images of nanoparticle assembly under DEP. Left: 1 MHz, 4 V, 1 min; Center: 1 MHz, 4 V, 2 min; Right: 1 MHz, 4 V, 4 min. Scale bars for all three images are 1 μm. The resistances corresponding to each device are 10 kΩ, 7 kΩ, and 7 kΩ, respectively.

resistance of this device was 7 kΩ, which is approximately the same as 2 min deposition experiments. It indicates that the resistance is likely constant, regardless of the quantity of nanoparticles between electrode tips, as the electrode junctions are connected by nanoparticles. This is different than earlier studies of DEP using 4-aminothiophenol-functionalized gold nanoparticles in nanoscale gaps. Under 1 MHz and 4 V DEP, the device resistances are reduced from 30 kΩ at 30 s to 10 kΩ at 3 min, and finally down to 3 kΩ at 10 min [6]. The variation in resistance pattern is likely due to the following reasons. First, the baseline resistance of a device is affected by the decay constant of the specific nanoparticle, which pertains to the chemical properties of the metal-ligand pair [7]. Second, the DEP trapping efficiency can be different from tip to tip, which can lead to a variation among resistances at individual junction, so the scaling up of resistance will be affected.

3.2. *Vapor Sensing Experiments*

The sensing experiments using oleylamine-functionalized palladium nanoparticles (OLA-PdNP) were performed at room temperature and pressure. Typical baseline resistances of sensors fabricated by OLA-PdNP were between 1 to 100 kΩ. Liquid analytes of isopropanol (IPA), acetone, hexane, cyclohexanone, acetonitrile, and dichloromethane (DCM) were delivered into 60-mL syringes and stored overnight to reach equilibrium. Saturated chemical vapors were mixed with dry air before being delivered to the sensor. The vapor concentration is reported as p/p_0, where p is the vapor partial pressure and p_0 is its saturation vapor pressure at room temperature. Sensor response is expressed as $\Delta R/R_0$, where R_0 is the baseline resistance and ΔR is the resistance change due to analyte vapor sorption at the nanoparticle film.

Sensor response profiles of eight devices measured simultaneously on the same chip are shown in Fig. 4. Each peak represents a detection signal from resistance change due to film swelling of the nanoparticles. The peak heights represent the sensor responses ($\Delta R/R_0$), which are between 1-6%. The peak width describes the dosing time of each pulse by syringe pump, which is set at 4 s. The chemiresistors are turned on by analyte vapors and their resistances increase, which indicates film swelling caused by the interaction between isopropanol and OLA-PdNP. The resistances drop back to the baseline once the dosing is stopped. Figure 5 illustrates the average responses of eight sensors measured simultaneously over the six chemicals. For most sensors, the highest signal is given by acetone vapor and the lowest signal corresponds to hexane vapor. The pattern of relative sensor signals can be interpreted as fingerprints that can be used to evaluate the robustness and accuracy of chemical vapor detection. The $\Delta R/R_0$ variation in different sensors for the same analyte indicates that sensitivity is related to variations in baseline resistance.

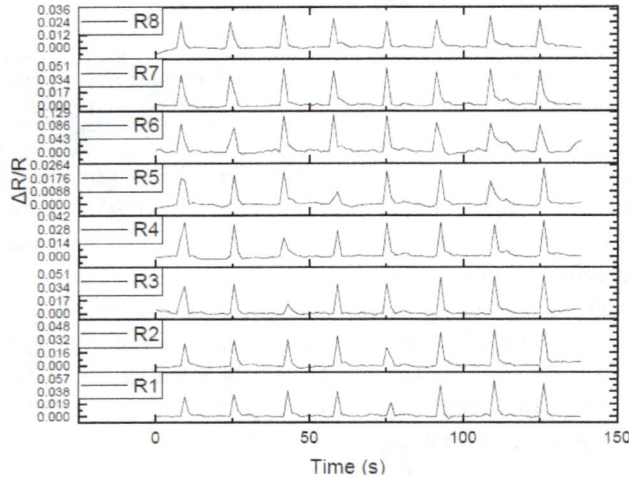

Fig. 4. Response profiles of eight sensors measured simultaneously to isopropanol vapors ($p/p_0 = 0.2$).

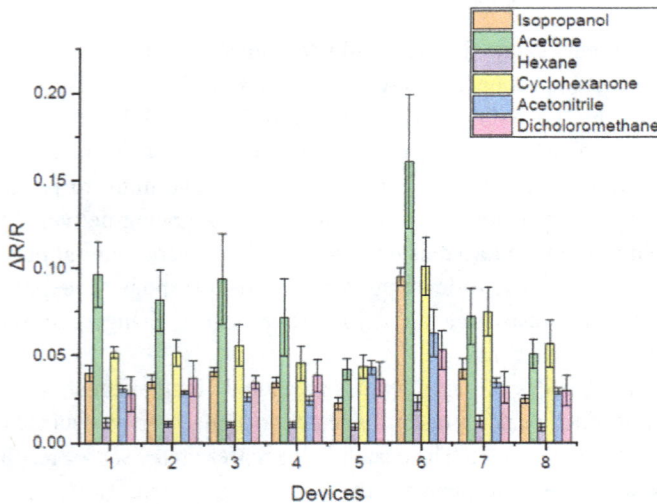

Fig. 5. Average responses of eight sensors on the same chip to six organic vapors ($p/p_0 = 0.2$). The error bars indicate the standard deviation over three replicate measurements.

3.3. *Pattern Recognition*

Common pattern recognition algorithm such as principal component analysis (PCA), linear discriminant analysis (LDA), and artificial neural networks (ANN) have been used to differentiate different chemical analytes or commercial products [8, 9]. Figure 6 shows 2-dimensional PCA score plots of six analytes using OLA-PdNP at $p/p_0 = 0.2$. Three replicates were performed in the vapor sensing experiment for each analyte over eight sensor devices on the same chip. The average $\Delta R/R_0$ of eight pulses in each replicate was used in the PCA matrix. Each data point on the score plot represents a replicate.

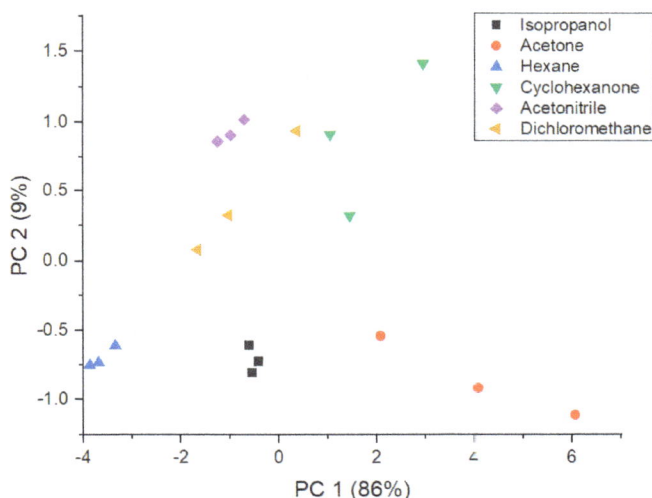

Fig. 6. PCA score plot of six analytes in eight sensors on the same chip at $p/p_0 = 0.2$. The first principal component has 86% variance and the second principal component has 9% variance. The sensing data were summarized into an 18 × 8 matrix. Each replicate is an observation, and each sensor corresponds to a variable.

The results from PCA score plot show two characteristics. Firstly, all chemical analytes are well separated with no overlaps. Among the six analytes, isopropanol, hexane, and acetonitrile form closely-bounded clusters, as compared to acetone, cyclohexanone, and dichloromethane, where data points are more scattered. This might be due to difference in partition coefficient among analytes. If the analyte stays longer in the sorbent phase, then the equilibrium of the vapor adsorption may be reached slowly so that the $\Delta R/R_0$ signal of the subsequent pulse will be affected. The mass and charge transport process can also lead to variations in precision of sensor performance. Secondly, the PCA score plot can reveal spatial information based on the chemical properties of analytes. For example, both acetone and cyclohexanone have a carbonyl group and their PC 1 score are both greater than 0. Meanwhile, acetonitrile (cyanide-containing) and dichloromethane (haloalkane) have a PC 2 score greater than 0, which is different than isopropanol and hexane.

4. Conclusions and Future Work

In this work, we demonstrated the use of oleylamine-functionalized palladium nanoparticles as chemiresistive sensing materials for fast and robust detection of organic vapors. Typical baseline resistance of 1-100 kΩ were achieved. Based on the results from nanoparticle assembly by dielectrophoresis and direct drop-casting, we concluded that the baseline resistances of a chemiresistor depend on the chemical properties of the metal-ligand pair, mainly affecting analyte vapor sorption and nanoparticle film swelling. The principal component analysis shows good separation of 6 common organic chemical vapors at a concentration of $p/p_0 = 0.2$.

The viability of incorporating functionalized-palladium nanoparticles onto many-devices on a single, miniaturized chip for robust and accurate discrimination of common organic vapors has been validated. By incorporating larger chemical versatility on a single chip, it may be possible to realize portable chemical sensors with improved accuracy and precision.

References

[1] F. J. Ibanez, and F. P. Zamborini, "Chemiresistive sensing with chemically modified metal and alloy nanoparticles," *Small*, vol. 8, 2012, 174-202.
[2] R. A. Potyrailo, "Toward high value sensing: monolayer-protected metal nanoparticles in multivariable gas and vapor sensors," *Chem. Soc. Rev.*, vol. 46, 2017, 5311-5346.
[3] M. Moreno, F. J. Ibanez, J. B. Jasinski, and F. P. Zamborini, "Hydrogen reactivity of palladium nanoparticles coated with mixed monolayers of alkyl thiols and alkyl amines for sensing and catalysis applications," *J. Am. Chem. Soc.*, vol. 133, 2011, 4389-4397.
[4] K. Fu, W. Pedrick, H. Wang, A. LaMarche, X. Jiang, B. G. Willis, S. Li, and Y. Wang, "Polynucleotide-functionalized gold nanoparticles as chemiresistive vapor sensing elements," *IEEE Sensors*, 2013.
[5] A. Ramos, H. Morgan, N. G. Green, and A. Castellanos, "AC electrokinetics: a review of forces in microelectrode structures," *J. Phys. D: Appl. Phys.*, vol. 31, 1998, 2338-2353.
[6] K. Fu, S. Chen, J. Zhao, and B. G. Willis, "Dielectrophoretic assembly of gold nanoparticles in nanoscale junctions for rapid, miniature chemiresistor vapor sensors," *ACS Sensors*, vol. 1, 2016, 444-450.
[7] N. Olichwer, A. Meyer, M. Yesilmen, and T. Vossmeyer, "Gold nanoparticle superlattices: correlating chemiresistive responses with analyte sorption and swelling," *J. Mater. Chem. C*, vol. 4, 2016, 8214-8225.
[8] E. Garcia-Berrios, T. Gao, J. C. Theriot, M. D. Woodka, B. S. Brunschwig, and N. S. Lewis, "Responses and discrimination performance of arrays of organothiol-capped Au nanoparticle chemiresistive vapor sensors," *J. Phys. Chem. C*, vol. 115, 2011, 6208-6217.
[9] R. Bro, and A. K. Smilde, "Principal component analysis," *Anal. Methods*, vol. 6, 2014, 2812-2831.

Threading Dislocations in Metamorphic Semiconductor Buffer Layers Containing Chirped Superlattices

Md Tanvirul Islam[1], Xinkang Chen[2], Tedi Kujofsa[3] and John E. Ayers[4]

Electrical and Computer Engineering Department,
371 Fairfield Way, Unit 4157, Storrs, CT 06269-4157, USA
[1]*md.t.islam@uconn.edu*
[2]*xinkang.chen@uconn.edu*
[3]*tedi.kujofsa@gmail.com*
[4]*john.ayers@uconn.edu*

Metamorphic realization of semiconductor devices has become increasingly important due to the great freedom it affords in layer compositions and thicknesses. However, metamorphic growth is often accompanied by the introduction of high densities of threading dislocation defects. This behavior may be understood by using an annihilation and coalescence model for the threading dislocation behavior which is based on the dislocation interaction length L_{int}. For its application we considered only glide of dislocations, so the interaction length was assumed to be equal to the length of misfit dislocation segments L_{MD}. The length of misfit segments was determined approximately by the Matthews, Mader, and Light model [*J. Appl. Phys.*, 41, 3800 (1970)] for lattice relaxation, and was assumed to be independent of the distance from the interface. Within this framework we have applied the annihilation and coalescence model to chirped semiconductor superlattices to evaluate these superlattices as strain-relaxed buffers for metamorphic devices. In this work we have studied two basic types of InGaAs/GaAs chirped superlattice buffers: type I superlattices are compositionally modulated while type II superlattices are thickness modulated.

Keywords: Chirped superlattices; graded buffer layers; dislocation interaction length; threading dislocations; semiconductor heterostructures and devices.

1. Introduction

Chirped superlattices are of interest as buffer layers in metamorphic semiconductor device structures such as high electron mobility transistors (HEMTs) and laser diodes, because they can combine the mismatch-accommodating properties of compositionally-graded layers with the dislocation filtering properties of superlattices. Here we compared the surface threading dislocations of compositionally-modulated and thickness-modulated InGaAs superlattice buffer layers on GaAs (001) substrates.

2. Methods

In this work we evaluated the average and surface threading dislocation density for two types of chirped superlattices, shown in Fig. 1, using a dislocation interaction length model for the annihilation and coalescence reactions between dislocations. The type I superlattice shown in 1(a) is compositionally modulated while the type II superlattice shown in 1(b) is thickness modulated.

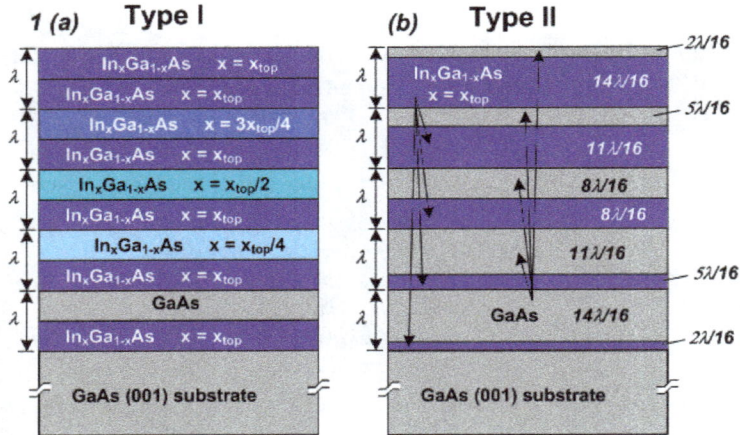

Fig. 1. Chirped InGaAs superlattice buffers considered in this work. The type I superlattice shown in (a) is compositionally modulated while the type II superlattice shown in (b) is thickness modulated.

In order to find the threading dislocation density profiles, the equilibrium strain and misfit dislocation density profiles were determined using an electric circuit model for semiconductor strained epitaxy [1]. Only glide of dislocations was considered, so the dislocation interactions length L_{int} was assumed to be equal to the length of misfit dislocation segments L_{MD}. The length of misfit segments was found in an approximate way using the Matthews, Mader and Light (MML) model for lattice relaxation [2], and was assumed to be independent of distance from the interface. Within the MML approximation, the lattice relaxation δ is given as a function of time t by

$$\delta = \beta[1 - e^{-\alpha t}] \tag{1}$$

where

$$\alpha = \frac{2Gb^3\rho_0(1 + v)cos\phi cos^2\lambda D_0 exp(-U/kT)}{(1 - v)kT} \tag{2}$$

and

$$\beta = f - \epsilon_{eq} , \tag{3}$$

where G is the shear modulus, b is the length of the Burgers vector, ρ_0 is the misfit dislocation density at the interface, v is the Poisson ratio, ϕ is the angle between the normal to the slip plane and that direction in the interface which is perpendicular to the intersection of the glide plane and the interface, λ is the angle between the Burgers vector and that direction in the interface which is perpendicular to the intersection of the glide plane and the interface, D_0 is the diffusion coefficient for a dislocation core, U is the activation energy for dislocation glide, k is the Boltzmann constant, T is the temperature, f is the lattice mismatch strain, and ϵ_{eq} is the equilibrium strain, found using the circuit model. In this work D_0 was estimated as 1.7 x 10^{-9} cm^2/s for In$_x$Ga$_{1-x}$As and assumed to be independent of the indium composition; the density of misfit dislocations ρ_0 was estimated as $\rho_0 = [|f|/b]^2$.

Using the MML model for lattice relaxation and the average lattice mismatch strain for each structure, the length of misfit dislocations L_{MD} was estimated by finding the effective stress τ_{eff}:

$$\tau_{eff} = \frac{2(\beta - \delta)G(1 + v)cos\phi cos\lambda}{(1 - v)} \tag{4}$$

The effective stress was used to determine the dislocation glide velocity,

$$v = B\tau_{eff}exp(-U/kT), \tag{5}$$

where B was assumed to be $6 \times 10^{-9}cm^3 dyn^{-1}s^{-1}$. The dislocation glide velocity was then integrated over time to find the approximate length of misfit dislocations in the structure:

$$L_{MD} = \int_{h_c/g}^{h/g} \frac{2GBcos\phi cos\lambda \beta exp(-\alpha t)exp(-U/kT)}{(1 - v)} dt, \tag{6}$$

where h_c is the critical thickness for the onset of lattice relaxation, h is the buffer layer thickness, and g is the growth rate, assumed to be $5\mu m/hr$.

The threading dislocation density profile was determined including interactions between misfit and threading dislocations as well as second-order interactions among threading dislocations, using the following equation [3]:

$$\frac{dD}{dz} = \frac{4\rho(z)}{L_{MD}sign \int_0^z \rho(\xi)d\xi} - L_{MD}D^2 \tag{7}$$

where D is the threading dislocation density, z is the distance from the interface, ρ is the areal density of misfit dislocations and L_{MD} is the length of misfit dislocations. D_0 was estimated as 1.7 x 10^{-9} cm^2/s for In$_x$Ga$_{1-x}$As and assumed to be independent of the indium composition. We assumed T = 700°C and used the material parameters for GaAs, as follows: b = 4 x 10^{-8} cm, v = 0.31, G = 32 x 10^{10} dyn/cm^2, U = 1.4 eV, λ = 60°, and ϕ = 35.26°.

3. Results and Discussion

Figure 2 shows the equilibrium in-plane strain profiles for representative type I and type II chirped superlattices having five periods, a total thickness of 200 nm, and a top indium composition of x_{top} = 0.4. Both types alternate between compression and tension, potentially allowing the filtering of dislocations by either the enhancement of coalescence and annihilation (the reactions between pairs of threading dislocations) or by dislocation compensation (the bending over of existing threading dislocations to create misfit dislocation segments).

In Figs. 3 and 4 we show the surface and average threading dislocation density (TDD), respectively, as functions of the total thickness and with top composition as a parameter, for type I and type II chirped superlattices. These characteristics show a general trend of lower TDD with greater thickness, as expected. However, the surface TDD in type II chirped superlattices shows unique behavior associated with dislocation compensation, which produces sharp dips in the surface TDD. Similar behavior is seen in Figs. 3 and 4,

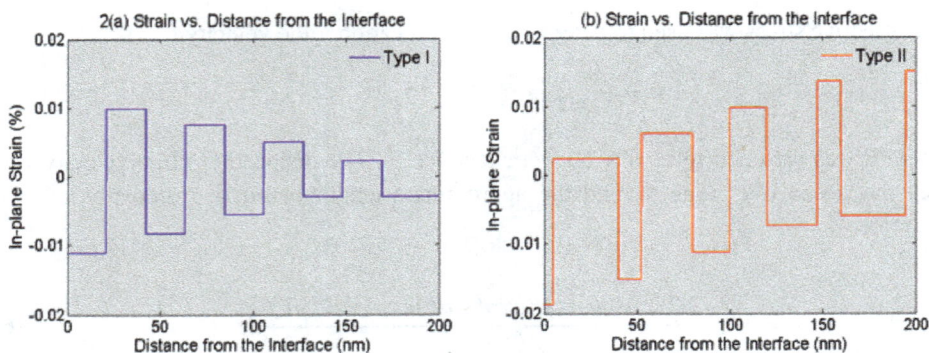

Fig. 2. Equilibrium strain profiles in representative type I (a) and type II (b) chirped superlattices having five periods, a total thickness of 200 nm, and a top indium concentration of x_{top} = 0.4.

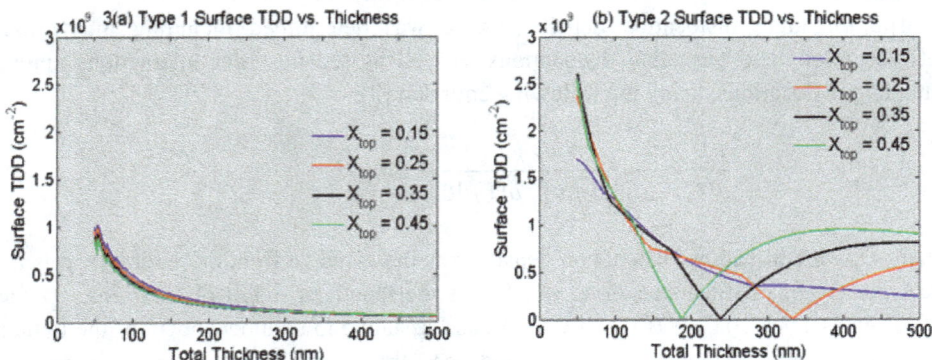

Fig. 3. Surface threading dislocation densities (TDD) in type I (a) and type II (b) chirped superlattices having five periods, as a function of total thickness and with the top indium concentration x_{top} as a parameter.

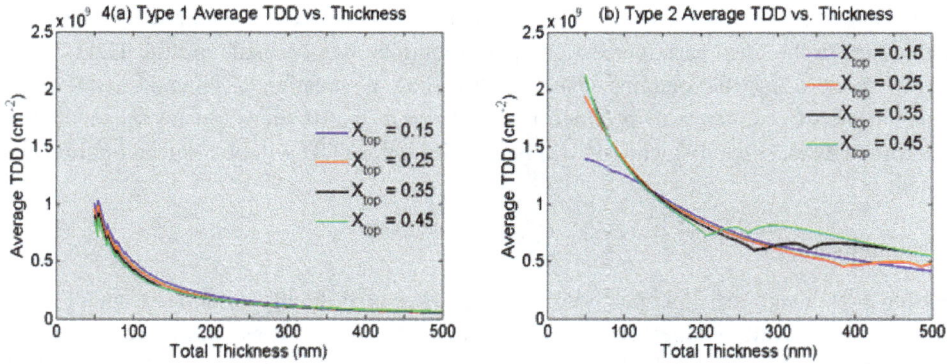

Fig. 4. Average threading dislocation densities (TDD) in type I (a) and type II (b) chirped superlattices having five periods, as a function of total thickness and with the top indium concentration x_{top} as a parameter.

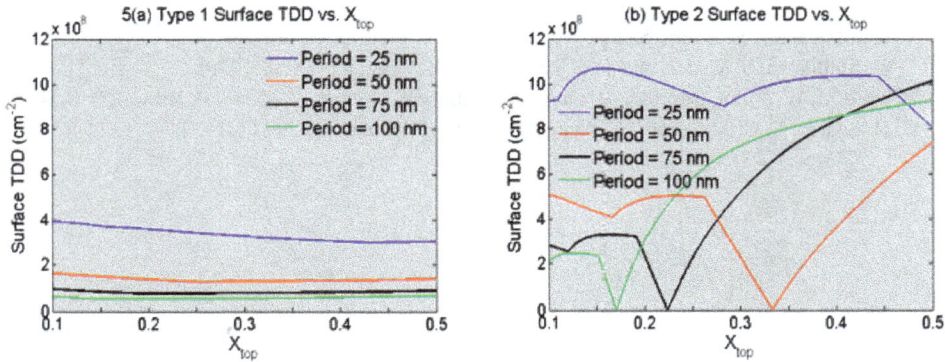

Fig. 5. Surface threading dislocation densities (TDD) in type I (a) and type II (b) chirped superlattices having five periods, as a function of the top indium concentration x_{top} with the period as a parameter.

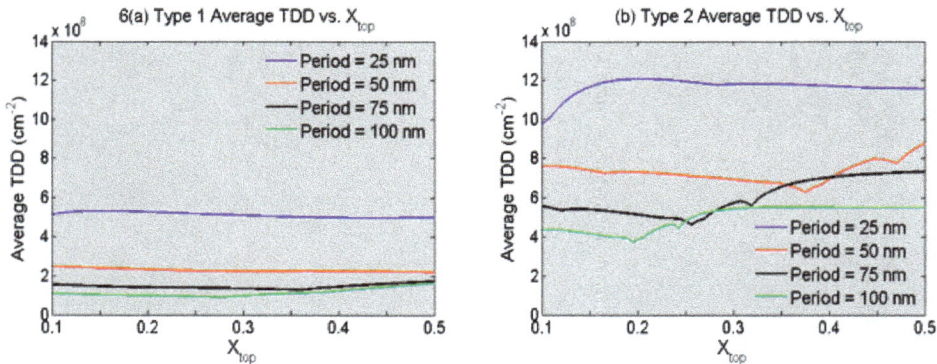

Fig. 6. Average threading dislocation densities (TDD) in type I (a) and type II (b) chirped superlattices having five periods, as a function of the top indium concentration x_{top} with the period as a parameter.

which show the surface and average TDD as a function of x_{top} with thickness as a parameter. In the ideal case, perfect compensation may be obtained, and the TDD can be driven to zero by the bending over of threading dislocation associated with misfit dislocations of one sense to produce misfit dislocations of the opposite sense. In real heterostructures, some dislocations will be sessile so the TDD will be reduced but may not vanish.

4. Conclusion

We have evaluated the TDD behavior in Type I (compositionally modulated) and Type II (thickness modulated) superlattice buffer layers. These two types exhibit very different behavior, and Type II superlattices allow complete compensation of the dislocations if all defects are glissile.

References

[1] T. Kujofsa and J. E. Ayers, *Semicond. Sci. Technol.*, 31, 115014 (2016).
[2] J. W. Matthews, S. Mader, and T. B. Light, *J. Appl. Phys.*, 41, 3800 (1970).
[3] T. Kujofsa, W. Yu, S. Cheruku, B. Outlaw, S. Xhurxhi, F. Obst, D. Sidoti, B. Bertoli, P. B. Rago, E. N. Suarez, F. C. Jain and J. E. Ayers, *J. Electron. Mater.*, 41, 2993 (2013).